Sustainability Programs

Sustainability Programs

A Design Guide to Achieving Financial, Social, and Environmental Performance

William Borges and John Grosskopf

Library of Congress Cataloging-in-Publication Data applied for

Hardback ISBN: 9781394307388

Cover image: Wiley
Cover design: © Moncherie/Getty Images

Set in 9.5/12.5pt STIXTwoText by Straive, Chennai, India

SKY10100320_031725

Contents

Climate change and other environmental and social impacts are real, pervasive, and increasing thereby affecting nearly every organization regardless of industry sector, size, or location.

Although current progress in managing sustainability issues is promising, it is typically reactive, piecemeal, and far too limited to environmental concerns. Increasingly, organization leaders recognize these shortcomings and seek innovative, financially beneficial responses to environmental impacts, social needs, and organizational governance. Plus, they are seeking better, cost-effective ways to anticipate and respond to pressures from equity holders, regulators, advocates, competitors, customers, and communities.

While promising individual sustainability initiatives abound, they have yet to be integrated into the broad body of conventional organizational management principles and practices. Because sustainability is relatively new and ill-defined, organizations often need help integrating it into their overarching *management system* (MS). As a result, sustainability initiatives are often *ad hoc* efforts and distractions that address *squeaky-wheel* and *flavor-of-the-month* issues. Consequently, current efforts in most organizations cannot achieve sustainability's promise and full value as an essential management discipline.

The Distinction Between Sustainability and Environmental, Social, and Governance

Although the terms *sustainability* and *environmental, social, and governance* (ESG) are often used interchangeably, it is essential to distinguish between them. Focused on ensuring the long-term viability of an organization's operations, sustainability is a holistic approach focused on managing the dynamics between environmental, social, and economic matters. The goals are to benefit financially while contributing to society and reducing adverse environmental effects. Complimenting this approach, ESG uses specific criteria to assess and communicate an organization's efforts to manage sustainability-related risks and opportunities. In simpler terms, sustainability can be considered the management activities dealing with social and environmental risks and opportunities, whereas ESG evaluates and reports those activities.

This distinction is significant because most of the book's organizational development prescriptions focus on creating systemic and systematic organizational structures and processes to identify and resolve specific environmental, social, and financial issues. However, the focus shifts in the last chapter, Chapter 11, to the overall sustainability program efficacy assessment and reporting activities that are essential to ESG evaluations and reports. Instead of a common approach, where an organization searches through its units, functions, and departments for miscellaneous *people, planet, and profit* (3P)-related activities to fill the pages of a periodic ESG-focused report, the book's approach provides an organization with intentional, substantial programmatic accomplishments.

Operationalizing Sustainability

Sustainability and ESG are undergoing far-reaching transformations due to widespread recognition of ecological degradations, social inequities, ever-increasing competitive and stakeholder pressures, and rapid changes in legislation and regulations. In response, this book systematically guides organizations in comprehensively identifying, prioritizing, and acting on their most pressing 3P needs within their resource limits and opportunities. Further, it provides them with proven management concepts and processes to operationalize sustainability in a manner that enables fact-based, greenwash-free ESG evaluations and reporting.

Unlike other approaches to sustainability program development that stress sustainability's specialized concerns and altruistic motivations, which, of course, are critically important, this book translates such ideas into these conventional management activities:

- Risk management
- Cost and expense reduction by eradicating wastes of all kinds
- Revenue enhancement through product and service innovation, and
- Competitive advantage gains through transparency.

By doing so, the business case for sustainability can be seamlessly integrated into organizational management activities, making it appealing to even the most ardent of critics.

Sustainability Management System Scalability

The book's *sustainability management system* (SMS) design, implementation, and management processes can best be realized in large-scale enterprises with approximately 2000 to 5000 employees. These typically have the critical mass of material and personnel resources, organizational agility, and management expertise to effectively employ the book's methods.

Irrespective of size, any organization in the private and public sectors with rudimentary technical capabilities and determined management can operationalize sustainability by selectively tailoring the book's ideas and processes. Selective tailoring is necessary to scale down the concepts and methods at smaller and resource-constrained organizations. Not to be left out are the huge, highly complex organizations that can benefit by scaling up the ideas and processes in this book, no matter the current status of their sustainability programs.

Achieving Sustainability with Familiar Management Concepts

The SMS design, implementation, and management processes are based on proven *continuous improvement* (CI) models used by private- and public-sector organizations. The origins and evolution of the book's SMS model are described in Appendix A. Underlying these processes is the *Plan-Do-Check-Act* (PDCA) *cycle,* an essential concept in well-known CI procedures such as Lean Manufacturing, Six Sigma, Total Quality Management, and the various International Organization for Standardization (ISO) MS standards.

Notably, by focusing on discrete quality, environmental management, health and safety, energy, and social responsibility topics, ISO and other standards are not often integrated into a single, more efficient MS that comprehensively addresses sustainability's 3P objectives. However, the newly released *ISO/UNDP PAS 53002 Guidelines for Contributing to the United Nations Sustainable Development Goals* significantly expand the range of MS sustainability topics. This book's SMS model provides a timely and highly effective method to integrate new sustainability and previously released standards and guidance into an organization's overarching MS. By integrating this wider range of sustainability-related concerns with general management ones, existing CI MSs have a better chance of achieving sustainability's full 3P potential.

The good news for organizations currently using topical CI MSs – especially those based on ISO standards – is that this book's methods require few fundamental modifications to achieve sustainability goals. The most significant change for these MSs is the addition of sustainability concerns to the current general management issues addressed during periodic needs assessments and performance improvement cycles.

For those organizations operating with conventional *management-by-objectives* (MBO) systems, the book's CI methods provide an effective SMS model for improving risk management, expense and cost reduction, revenue enhancement, and tangible and intangible competitiveness.

The Book's Approach to Sustainability Program Management

This how-to book provides a detailed step-by-step PDCA-based process to create a no-compromise SMS. An SMS links – or, some say, hardwires – an organization's sustainability policy and strategic intentions to its day-to-day administration and operational activities. Operationalizing sustainability in any organization involves these foundational management activities:

- Formally deciding to create a sustainability program
- Developing an overarching sustainability policy
- Designing organizational structures and processes to implement the policy
- Periodically defining, prioritizing, and shortlisting the organization's most pressing sustainability needs, and then cascading capability creation and performance improvement accountabilities down through the organization
- Designing, tracking, and evaluating projects and other initiatives to achieve those accountabilities with measurable benefits that exceed costs, and
- Using lessons learned to continually improve the sustainability program.

Critically, the book addresses an organizational transformation issue that many, if not most, technical practitioners and organizational leaders need to address better: formal change management. Experienced change agents know that the root causes of underperformance and failure of many initiatives are inadequate planning and irresolute execution. This is especially true for complex, comprehensive sustainability programs. Aspiring change agents can familiarize themselves with the information in Appendix B, *Essential Change Management Concepts.*

The Authors' Qualifications

The authors have helped private- and public-sector organizations of all sizes throughout the United States and abroad for over four decades. They are sustainability pioneers who have spent much of their technical and management careers using methodical CI processes to solve intractable problems and create benchmark MSs. Further, they have developed innovative curricula and taught and mentored hundreds of students and professionals in systematic performance improvement and sustainability principles and practices. They have also published on sustainability, environmental, health and safety, quality, engineering, security, and business excellence topics. In addition to speaking at numerous conferences and other forums throughout the United States and abroad, they founded and led a nonprofit dedicated to sustainability education for private- and public-sector organizations in southern California. Though never seeking them, the authors and the organizations they represented have received numerous awards, distinctions, and recognitions. The author's biographies are included at the end of the book.

Acknowledgments

We extend our gratitude to everyone who has contributed to this book.

Special thanks go to John Milliman, PhD, Aisha Cissna-Cervantes, Kirsten Anderson, Salem Afeworki, Oumaima ben Amor, and Cheryl McCorkle, who generously donated their time and efforts to produce clear-eyed reviews of early drafts. They provided much-needed perspectives on the needs of Gen Z to Baby Boom readers and aspiring-to-seasoned sustainability and management professionals.

We are incredibly grateful for all our friends and colleagues who have been enthusiastic supporters. Although there are so many to name, we would be remiss if we didn't include our long-time colleagues and supporters, without whose encouragement this book would not have been written: Ozzie Paez, Stephen Evanoff, Al Hurt, Michael Mutnan, Laura Ann Hopeman, Paolo Espaldon, Peter Burgess, and Federico Fioretto … *all experts in their fields.*

We also wish to recognize our past colleagues, clients, and students who, for over four decades, shaped our experiences and views on sustainability and its environmental, social, health and safety, quality, and business excellence topics. Your work, whether at ISONetwork, General Dynamics, ADAC Laboratories, USMC Base Camp Pendleton, Renown Health, or Cuyamaca and MiraCosta Community Colleges, has had a profound impact. Who knew the effect our ahead-of-their-time efforts would have? Working with you made us better, which made this book better.

Most of all, we thank our ever-patient and supportive wives, Pearl Grosskopf and Rosalind Borges. We promise we will get back to our other chores … *honestly!*

Acronyms

3P	People, Planet, and Profit
3R	Reduce, Reuse, Recycle
5W's & 1H	What, Why, Who, Where, When, and How
ASTM	American Society for Testing and Materials
B2B	Business to Business
BCEE	Board Certified Environmental Engineer
BSCE	Bachelor of Science in Civil Engineering
CERES	Coalition for Environmentally Responsible Economies
CEO	Chief Executive Officer
CFO	Chief Financial Officer
CI	Continuous Improvement
CM	Change Management
CSDDD	European Union Corporate Sustainable Due Diligence Directive
CSR	Corporate Social Responsibility
CxO	Any Type of Chief Corporate Officer
EMS	Environmental Management System
ERP	Enterprise Resource Planning
ESG	Environmental, Social, and Governance
FMEA	Failure Modes and Effects Analysis
GD	General Dynamics Corporation
GRI	Global Reporting Initiative
H&Q	Hambrecht and Quist Investment Bank
IFRS	International Financial Reporting Standards
IPO	Initial Public Offering
ISSB	International Sustainability Standards Board
ISO	International Organization for Standardization
KISS	Keep It Simple, Stupid
KPI	Key Performance Indicator

KT	Knowledge Transfer
LCA	Life Cycle Assessment
MBA	Master of Business Administration
MBO	Management by Objectives
MDC	Monitor Detect Correct
MS	Management System
NASA	National Aeronautics and Space Administration
NGO	Nongovernmental Organization
OD	Organizational Development
OFI	Opportunity for Improvement
PDCA	Plan-Do-Check-Act
PM	Project Management
R&QA	NASA's Reliability and Quality Assurance Office
RBA	Responsible Business Alliance
RCA	Root Cause Analysis
ROI	Return on Investment
SASB	Sustainability Accounting Standards Board
SEC	Securities and Exchange Commission
SMART	Specific, Measurable, Achievable, Relevant, and Time-Constrained
SME	Small- and Medium-Size Enterprises
SMS	Sustainability Management System
SST	Sustainability Specialty Team
STEM	Science, Technology, Engineering, and Mathematics
SWOT	Strength, Weakness, Opportunity, and Threat Analysis
T&D	Training and Development
TANSTAAFL	There Ain't No Such Thing as a Free Lunch
TQM	Total Quality Management
WBS	Work Breakdown Structure
WIIFM	What's in It for Me?
UNSDG	United Nations Sustainable Development Goals

1

Introduction

There are many ways to define sustainability. One of the most elegantly simple ones is the *United Nations' 1987 Brundtland Commission's* definition wherein the global perspective of sustainable development is:

> *Meeting the needs of the present without compromising the ability of future generations to meet their own needs.*

Of course, the definition's elegance conceals the complex difficulties in applying high-level sustainability concepts to an individual organization's specific activities. In organizational management terms, it can be defined this way:

> *Throughout a product or service's entire closed-loop life cycle, sustainability is how an organization creates value by maximizing its activities' positive social, environmental, and economic effects while eliminating or minimizing their adverse effects.*

It has taken decades for innovators and early adopters to develop efficient technical and management methods to achieve sustainability goals. Building on these efforts, this book provides organizations of various types and sizes with a practical design, implementation, and management model for a *continuous improvement* (CI) *SMS*.

SMSs focused on measurable 3P performance are one of sustainability's most innovative and effective best management practices. The authors are enthusiastic advocates of the SMS concept, having successfully initiated, designed, implemented, and managed such systems throughout their careers. With this experience, they have incorporated innovators' and early adopters' methods and

Sustainability Programs: A Design Guide to Achieving Financial, Social, and Environmental Performance, First Edition. William Borges and John Grosskopf.
© 2025 John Wiley & Sons, Inc. Published 2025 by John Wiley & Sons, Inc.

success factors – along with their own – into this book's ever-evolving model. To share this aggregated knowledge and experience, the book prescribes:

An adaptable stepwise model to create an organization-wide, continuous-improvement management system to achieve sustainability goals with the least cost, effort, and risk.

The origins and evolution of the book's SMS model are described in Appendix A.

1.1 SMS Best Management Concepts

Despite the growing awareness of climate change, widespread environmental degradation, resource depletion, supply chain disruptions, and social ills over the past half-century, it has taken this long for a relatively few organizations to develop effective best management practices to systematically identify, prioritize, and act on their most pressing 3P needs with the least cost, effort, and risk. With innovators and early adopters providing promising solutions, it is time to share with – and encourage their widespread adoption by – other private and public organizations.

Recognizing that these best management concepts were developed in the private sector over years of trial and error is essential. Because its activities collectively and directly cause the most consequential environmental and social impacts – good and bad – the private sector must, out of necessity, lead the transition to a more sustainable economy. Unlike public and nonprofit sectors, it has the resources, expertise, and motivation at micro- and macroeconomic scales required for rapid transformational change via the profit motive.

Integral best management practices in SMSs include:

- Policy-initiated mandates implemented through needs-assessed sustainability strategies, tactics, and initiatives cascaded throughout organizations using work processes based on the PDCA cycle, and
- High degrees of proactivity, collaboration, and transparency.

The second point cannot be overstated. The causes of climate change and other environmental and social impacts are so broad and complex that they defy simple in-house solutions. As a result, proactive private sector organizations have learned and now promote

Unless someone like you cares a whole awful lot, nothing is going to get better. It's not.

**Dr. Seuss/T.S. Geisel
The Lorax**

the idea that individual organizations cannot go it alone in becoming sustainable. They must collaborate on this transformation with all other economic, social, technical, and government sectors.

1.2   Target Readers

The target readers for this book are:

- Private sector board members, chief corporate officers, and other executives and managers, along with elected officials and executive directors at public sector and nonprofit organizations
- Sustainability champions and supporters at all organizational levels
- Technical and management professionals with responsibilities such as sustainability, environment, health and safety, and corporate social responsibility, and
- College and university educators, training-and-development professionals, and their students.

At its most elemental level, this book is an *organizational development* (OD) handbook written to help sustainability change agents in their transformation efforts, whatever their title and position. Who is a change agent?

> *A change agent is a person who, by personal commitment or formal assignment, takes on the role of leader to champion the initiation, design, implementation, and successful management of a transformational idea.*

Further, this book can guide those persons throughout an organization who support change agents in their efforts. Such widespread involvement and support are essential to achieving an organization's sustainability intentions.

Readers are assumed to be familiar with basic organizational management concepts and current sustainability issues. However, it is recognized that this will not always be the case, especially for technically oriented sustainability professionals early in their careers and managers and business specialists who focus exclusively on conventional organizational performance. The most rudimentary management concepts are often unfamiliar to many technical professionals. After all, when could they attend formal management courses during their technically packed academic schedules? Likewise, organizational leaders and personnel for whom sustainability is still an ill-defined esoteric concept have a comparable problem. When could they take additional physical, biological, or social science courses in their administration or operations specialty curricula?

This book does not attempt to address these education shortcomings. However, to help readers needing to broaden and deepen their knowledge and skill sets, Appendix C, *Sustainability Learning and Information Resources*, provides lists of sustainability-themed masters of business administration degree programs, professional certificate and accreditation programs, specialty short courses, and other resources.

1.3 The Book's Objectives

Companies must prepare for and actively participate in the rapid evolution of *environmental, social, and governance* (ESG) and sustainability laws, regulations, directives, and standards. To this end, the book seeks to guide aspiring and current sustainability leaders in any organization in developing an SMS tailored to its unique needs. Its focus is the systematic – and systemic – application of sustainability principles and practices in any organization. The book's objectives are to:

- Define sustainability in terms of its organizational performance benefits
- Examine requisite management and technical principles and practices
- Prescribe a highly adaptable, stepwise process for the design, implementation, control, and improvement of a practical SMS that is fully integrated into an organization-wide MS
- Offer examples of exemplary sustainability programs, publications, and university- and practitioner-level curricula, and
- Stress urgency in producing measurable sustainability results throughout value chains and other stakeholder communities.

1.4 The Need for This Book

As mentioned in the Preface, although an increasing number of private and public sector organizations have embraced sustainability over the past few decades, large-scale progress has been limited, piecemeal, and too often restricted to environmental issues.

Excellent firms don't believe in excellence – only in constant improvement and constant change.

Tom Peters

There are several reasons that current efforts are less widespread and effective:

- Despite the successes of innovators and early adopters, sustainability has not yet matured into a universally accepted body of management principles and practices that can be readily implemented within various organization types, sizes, and locations
- With their typical emphasis on technical disciplines and academic research, educational institutions are not keeping up with the rapid evolution in sustainability's organizational transformation practices
- There is an overreliance on topic-limited performance standards that do not comprehensively address sustainability across its full 3P spectrum
- There are powerful industry, financial, and political forces working to thwart sustainability's widespread adoption, and

- Advocates and various media have been less than effective in communicating the benefits of sustainability.

Many leaders fail to understand sustainability's potential to improve all aspects of organizational performance throughout a value chain, especially in the areas of:

- Reduced risks
- Reduced costs and expenses, especially those associated with energy consumption and waste generation
- Increased revenues through product and service innovation, and
- Enhanced tangible and intangible competitive advantages through transparency.

1.4.1 Helping Leaders Embrace Sustainability

Resolving the problem listed earlier – i.e. effectively communicating sustainability's benefits – is challenging because it requires helping reluctant and recalcitrant leaders understand that sustainability is not a burden – it is the opposite, a benefit – when treated as an advanced approach to overall performance improvement. As such, it provides organizations with new management insights and methods for adding and creating value throughout the life cycles of their product and service value chains.

Fortunately, there are now enough examples of positive sustainability programs – as well as ESG reporting pressures from the financial and government sectors – that many laggard leaders realize they are falling behind their competitors in domestic and international markets. This book's prescriptions may be the best path forward for them.

1.4.2 The Business-as-Usual Problem

However, despite clear evidence that fundamental change is required, *business-as-usual* remains the norm for many organizations. Business-as-usual blinds leaders to opportunities to grow organizations in new, sustainable ways. Further, business-as-usual creates unnecessary sustainability-related risks to organizational performance due to rapidly changing environmental, social, economic, and regulatory conditions.

> *The largest room in the world is the room for improvement.*
>
> **Helmut Schmidt**

1.4.3 Change is Coming Ready or Not

The evolving ESG discipline originated as a financial concept intended to protect investors, address stakeholder sustainability concerns, and quantify sustainability impacts and risks. Rapidly changing regulatory conditions are crucial concerns for companies operating in the European Union, where sustainability reporting requirements have been introduced and are quickly evolving. Further, for large domestic publicly traded companies that have yet to report on sustainability performance voluntarily, the United States Securities and Exchange Commission (SEC) has now mandated it. The new requirements include reporting sustainability-related corporate goals and how those goals will be achieved. Such game-changing reporting requirements will be the tail that wags the dog for companies needing to adopt structured approaches for creating and completing successful sustainability initiatives. As a result, business-as-usual days may be over for these companies. Fortunately, though, this book provides an OD pathway to:

Business as <u>unusual</u>.

1.4.4 Stumbling Blocks in the Way of Creating a Sustainable Organization

The big question for leaders assuming responsibility for creating a sustainability program remains, though:

How do you do it?

Follow-on questions inevitably include:

How much will it cost?
And
How long will it take?

Dialogues at conferences and in industry press, discussion groups, blogs, and other media swirling around the creation and management of sustainability programs are fraught with superficial efforts, quick fixes, misdirections, conceptual and buzzword overloads, and – to be blunt – a lot of OD nonsense. Instead of making sense of the racket, many sustainability change agents withdraw to what they know best. In doing so, they understandably approach sustainability impulsively from their narrow perspectives, skill sets, and experiences. These skill sets and experiences are frequently focused on technicalities or inconsequential distractions that have little to do with an organization's most pressing sustainability needs.

Such focuses typically result in *ad hoc* projects responding to *squeaky-wheel* or *flavor-of-the-month* issues. These projects are not bad in themselves. After all, they may resolve one or more issues and demonstrate that sustainability improves organizational performance. However, depending on the professional background of the person leading such efforts, characteristic problems may arise in their development.

- A typical inclination for technical professionals is to jump in and start working on whatever type of project is most familiar, easily accomplished, or budget-friendly.
- The recurring inclination for organizational managers is to do a public relations initiative or a trendy small-scale cost-saving project. Regretfully, many of these efforts risk backfiring as ineffectual and costly greenwash, sustainability's cardinal sin.

There is nothing inherently wrong with these *quick-win* impulses as long as there is no expectation of transformational change in how an organization conducts business. Such efforts can help build collective confidence that sustainability is good for business.

However, such *ready-FIRE-aim* efforts risk failing to identify and meet the organization's most pressing needs while wasting scarce financial, labor, material, temporal, and tangible and intangible marketplace goodwill resources. It's the kind of shortsighted thinking succinctly expressed in this quote from H. L. Mencken:

For every complex problem, there is an answer that is clear, simple, and wrong.

1.4.5 The Critical Importance of Sustainability Program Focus

Most importantly, such efforts risk failure to identify and meet the organization's most pressing sustainability needs. As noted earlier, this failure wastes scarce financial, labor, material, and temporal resources while squandering tangible and intangible competitive advantages and opportunities. It is because of this genuine risk that the methods in this book are needed to systematically and systemically:

- Formally decide to create a sustainability program
- Set an unambiguous sustainability policy
- Define, prioritize, and shortlist an organization's most pressing sustainability needs
- Meet those shortlisted needs within the organization's resource limits and opportunities with the least cost, effort, and risk, and
- Produce significant, measurable benefits that exceed costs.

1.5 How the Book's Approach Differs from Others

This book differs from other approaches with its adaptable step-by-step organizational design, implementation, and management process. The major steps are illustrated in Figure 1.1's generalized concept map. Notice that the steps are color-coded to show how the PDCA elements are related to the individual steps. Because the *do* and *check* steps are so intertwined, they are joined together in the figure. The interrelationships between the steps in Figure 1.1 are detailed further in Chapter 2 and Figure 2.4's process map.

- The process starts by addressing the all-important formal *go/no-go* decision to create a sustainability program. For those organizations with existing loosely defined and controlled sustainability programs, this and subsequent steps need to be addressed retroactively. The crucial go/no-go decision requires consideration of tangible and intangible environmental, social, and other administrative and operational risks and benefits at the highest levels.
- With a *go-decision*, the process moves on to define an organization's sustainability policies.

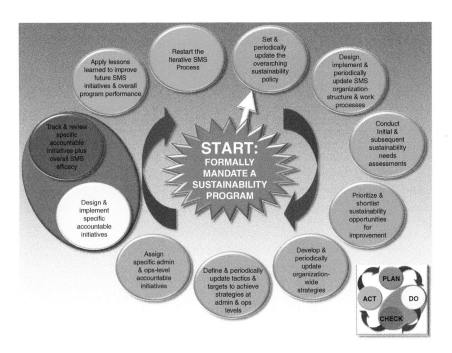

Figure 1.1 Generalized Concept Map: The SMS Model.
Source: W. Borges.

- Then, the process designs the organization structure and work activities needed to implement the sustainability policies.
- The process incorporates the idea that financial *return on investment* (ROI) analyses are an essential starting point, not the end point. Sustainability adds social and environmental considerations to ROI analyses.
- From there, the process prescribes how to:
  - Define and prioritize the organization's most pressing sustainability needs
  - Set overall time-limited strategic goals along with specific tactical objectives and targets cascaded into all units, functions, and departments
  - Design and complete sustainability-related capability creation and performance improvement initiatives with appropriate ROIs at all organization levels.
- Finally, the process improves the SMS structures, procedures, and activities with corrective actions and lessons learned before restarting the PDCA-based improvement cycle.

1.5.1 An Essential Concept: The PDCA Cycle

Based on the familiar PDCA cycle, an SMS has a distinct starting point, i.e. initial planning. However, cyclical SMSs have no defined end points. This is the PDCA cycle's fundamental strength:

> *When effectively applied, PDCA continuously and endlessly improves the organization's performance.*

1.5.2 An Emphasis on Program Evolution

A key concept in this book is:

> *Perfect should not get in the way of good enough.*

This is because, over time, CI methods derived from the PDCA concept take *good enough* to the next level by iteratively eradicating work process inefficiencies. Awareness of this concept helps minimize angst among leaders and staff when problems arise during sustainability program development.

1.5.3 An Emphasis on Change Management

Another departure from convention in this book's approach is its emphasis on formal CM principles and practices. Far too many organizations neglect CM at their peril. Ignoring the need to plan or not adequately planning for change is a chief cause of sustainability program underperformance or failure.

Initially, most leaders do not expect a sustainability program to transform how an organization operates. However, that is precisely what sustainability principles and practices have the potential to do through:

- New methods to *plan*, *organize*, *control*, and *lead* an organization
- Fewer risks, liabilities, and sanctions throughout the life cycles of product and service delivery value chains
- Lower administrative expenses and operational costs throughout the life cycles of product and service value chains by eradicating all kinds of waste
- Revenue streams enhanced through product and service innovation, and
- Enhanced tangible and intangible competitive advantages through transparency.

Although management initiates the change process, transformational change begins at the personal level and then extends to entire organizations. From there, individual organizations influence industries, the public sector, and society. Because of its importance, there are specific discussions about CM issues throughout the book, especially in Chapter 3 and Appendix B.

> *There is nothing more difficult to plan, more doubtful of success, nor more dangerous to manage than the creation of a new system. For the initiator has the enmity of all who would profit from the preservation of the old institutions, and merely lukewarm defenders in those who should gain by the new ones.*
>
> **Niccolo Machiavelli**

1.5.4 Emphasis on Adaptability

One of the most important lessons the authors have learned while initiating, designing, implementing, and running technical programs and MSs is:

> One size does not fit all.

Regrettably, for every large enterprise with the resources and expertise to build an effective SMS, far too many smaller organizations do not. Irrespective of resource capacities and constraints, this book provides the concepts and methods that organizations of any size can use to produce significant environmental stewardship, social responsibility, and financial results throughout their value chains. Adapting these concepts and techniques will help leaders customize their organization's sustainability program to meet their unique needs.

1.5.5 Emphasis on Integration

Finally, the book's adaptable SMS design, implementation, and management processes enable easy integration into an organization's current overarching

MS. Integration is a critically important efficiency tactic to help transform an organization into a greener and more socially responsible enterprise with the least cost, effort, and risk while improving productivity, overall financial performance, and stakeholder satisfaction.

1.6 When Can Measurable Progress Be Expected?

Sustainability and other similarly designed CI MSs often produce dramatic results by significantly reducing risks and other inefficiencies. The *Further Reading* examples provided throughout the book highlight the successful efforts of early SMS innovators and adopters.

It is essential to recognize that SMSs are *strategic initiatives*, i.e.:

> *A means by which an organization translates its goals into practice.*

Like most strategic initiatives, SMSs take time to achieve the intended results. There are no shortcuts. An applicable management adage is:

> *Work impatiently to achieve the intended results, but be patient while waiting for them.*

As strategic initiatives, SMSs are major transformations requiring leadership determination and highly focused CM planning and actions. The approach outlined earlier takes considerable effort and time to complete the initial OD tasks:

In order to go fast, you must first go slow.

Bob Willard

- Reach a go-decision
- Develop organization-wide sustainability policies
- Create an SMS management framework and associated work processes
- Identify and successfully instruct SMS management and technical personnel, and
- Conduct needs assessments.

Initial results have typically been slow for innovators and early adopter organizations that have created effective sustainability programs. However, with time and the PDCA cycle concept, the programs have become increasingly effective in producing measurable – and even dramatic – 3P results. Of course, the time required to achieve results will vary by organization. It depends on such factors as:

There are four purposes of improvement: easier, better, faster, and cheaper.

Shigeo Shingo

- The nature and urgency of sustainability challenges facing an organization
- Leadership determination
- Resource availability, and
- Efficacy of capability creation, performance improvement, and CM efforts.

Significant, measurable results can be produced early in an SMS's first year of operation. Early quick-win results to resolve important sustainability issues are encouraged to gain support for the SMS effort. In subsequent operating years, the SMS will become even more capable of producing remarkable results, even breakthrough performance, through its CI processes.

1.7 Notes about Industry Standards and Government Regulations

While determining how best to manage sustainability issues, proactive organizations have adopted and benefited from various industry standards and government regulations. This has provided them with valuable perspectives on integrating sustainability systematically and systemically into the range of conventional organizational management concerns.

The following are some of the more important standards and regulations affecting the design and operation of SMSs.

- High-level standards for sustainability management, including the:
  - *United Nations Sustainable Development Goals* (SDGs) and
  - *CERES Principles Code of Corporate Environmental Ideals.*
- Voluntary sustainability and financial performance reporting standards, including the:
  - *Carbon Disclosure Project* (CDP)
  - *Global Reporting Initiative* (GRI), and
  - *Sustainability Accounting Standards Board* (SASB) *Standards and Integrated Reporting Framework* administered by the *International Financial Reporting Standards* (IFRS) *Foundation's International Sustainability Standards Board* (ISSB).
- Basic MS standards and codified performance improvement methods, including the:
  - International Organization for Standardization (ISO) MS standards, especially *Guidelines for Contributing to the United Nations Sustainable Development Goals, ISO/UNDP PAS 53002*
  - Social Accountability International *SA8000 Social Certification Standard*
  - *Baldrige Performance Excellence Program*, and

- Process-oriented methodologies, including *Lean Manufacturing*, *Six Sigma*, and *Total Quality Management* (TQM).
- Various industry standards, such as the:
  - *Responsible Business Alliance (RBA) Code of Conduct*
  - *Marine Stewardship Council's Principles and Criteria for Sustainable Fishing*
  - *Sustainable Apparel Coalition's Higg Index*, and
  - *ASTM International's* and *ISO's* industry-specific technical standards.
- Individual organizations have also set standards for their suppliers, such as the *Walmart Sustainability Index*. The Index enables the company's buyers to evaluate supplier sustainability performance across the closed-loop life cycles of their products and services.
- Lastly, there are international, federal, state and local treaties, agreements, laws, regulations, and rules governing business. Compliance with them is crucial. As noted earlier, these include the game-changing:
  - *European Union's 2023 Corporate Sustainability Due Diligence Directive* (CSDDD), and
  - *United States SEC Final Rule on The Enhancement and Standardization of Climate-Related Disclosures for Investors.*

None of these standards provide an organization with a ready-to-implement SMS design. This is because they are not organizational design models. Instead, they are format and topical content standards that address *what* questions rather than the critical and more difficult organizational development *how* ones. This includes the new ISO *Guidelines for Contributing to the United Nations Sustainable Development Goals, ISO/UNDP PAS 53002*. Although these guidelines provide the most comprehensive outline of organizational concerns for an SDG-focused sustainability program to date, they too are limited to *what* questions. The topic-specific nature of the other relevant ISO MS standards and guidance – including quality, environment, energy, health and safety, and corporate social responsibility – are also restricted in their program design and implementation capabilities.

Despite their shortcomings, standards and regulations offer important SMS design considerations. Their underlying planning, organizing, controlling, and leading concepts are essential in designing and implementing an effective sustainability program that can fully integrate into an organization's overarching MS.

1.8 A Risk to Sustainability Program Success

As noted earlier, a significant risk has doomed many sustainability programs at their start: *greenwashing*.

Greenwashing uses environmental imagery, misleading language, and hidden trade-offs where an organization emphasizes a sustainable aspect of products or services while engaging in other environmentally damaging practices.

It is a rampant problem that damages reputations, risks regulatory and judicial sanctions, and ultimately threatens financial performance for organizations that knowingly or unwittingly employ the practice. Potential greenwash missteps in ESG reporting in the European Union and the United States compound this risk. This book's SMS design and operations prescriptions provide practical ways to immunize an organization against such blunders.

1.9 Consequences of Failing to Act

This book is focused on environmental, social, and financial performance through the ongoing reduction of risks and the exploitation of opportunities by cascading an organization's strategic sustainability intentions into the activities of every administration and operations unit, function and department. With solid management commitment, the book's prescriptions provide the means to transform a middling organization struggling or underperforming with sustainability principles and practices into a thriving one.

What happens to organizations that fail to fully embrace the sustainability concepts and processes outlined in the book? For smaller organizations and those with minor environmental and social footprints, the initial problems may seem insignificant. However, over time, the cumulative effects of missed opportunities to decrease risks, gain efficiencies, increase revenues, and achieve competitive advantages will become apparent. This is especially true for larger profit-margin-constrained organizations grappling with significant environmental and social issues.

The consequences for organizations that ignore the book's prescriptions altogether can be severe. Their marketplace adversaries will exploit their inaction, losing competitive advantages, key talent, customers, stakeholder goodwill, and positive financial performance. These losses may not be immediate, but they are inevitable. The time to act is now to avoid being left behind.

1.10 Advice for Leaders, Small Organizations, and Aspiring Sustainability Practitioners

Individuals and entire organizations can quickly become discouraged by the effort and time required to design, implement, and manage an effective sustainability program. It is a strategic process that requires considerable focus and determination to address newly discovered challenges effectively. This is particularly true for individuals and smaller organizations just starting. It also applies to those who have tried to create an effective program but faced setbacks or even discontinued their efforts.

Talk to those who run successful sustainability programs. They will tell you it is not only worth it, and many will add that they should have started long ago. *So, don't be discouraged!* The SMS model's practical and adaptable step-by-step methods can inspire newcomers to pursue their organization's sustainability goals and help floundering programs resolve their shortcomings. Significantly, they can accelerate the reader's and organization's learning curves.

1.10.1 There is No Such Thing as Overnight Success

Sustainability programs are strategic transformational efforts. Although beneficial results can be achieved early, more substantial gains may take several fiscal quarters or even years. This is especially true in large, complex organizations. However, the benefits never end thanks to CI principles and practices.

1.10.2 Sustainability Transformations are a Team Sport

Not only are they strategic, but the most effective sustainability transformations involve every business unit, function, department, and person. Some have more responsibilities than others, but each must know how they will contribute through their assigned roles and accountabilities. Importantly, every entity and person must act with purpose, determination, and urgency. The required systematic effort is analogous to a full-court press in basketball, a best practice in both sports and the creation of successful sustainability programs.

Having worked as an enterprise risk management program manager in two major General Dynamics divisions, co-author Grosskopf was promoted to the corporate level, where he developed a strategic perspective on and systems-based skill sets for organizational transformation. Likewise, co-author Borges developed his strategic leadership outlook by designing and implementing systems-based transformations at a Silicon Valley company, a start-up environmental engineering firm, a major military base, and a regional healthcare system. The lesson learned for both was that one person cannot effect change alone; successful change requires universal recognition of its need and enthusiastic commitment by all.

1.10.3 Learn

Many readers will realize they and their associates lack ideal sustainability-program skill sets. As suggested throughout the book, deficiencies should not be viewed as problems but as opportunities for growth and development. This is the perfect starting point for those requiring new skills to research the knowledge transfer offerings in Appendix C, *Sustainability Learning and Information Resources*. The appendix provides several learning resource options in the following categories:

- Sustainability-themed Master of Business Administration degree programs
- Sustainability-themed professional certificate and accreditation programs
  - University certificate programs
  - General certifications
  - Reporting certifications
  - Green building certifications
  - Operations in built environments certifications
  - ESG and sustainable finance certifications
  - Governance and risk certifications
  - Urban and infrastructure certifications
  - Other specialty certifications
- Sustainability-themed short courses and other resources
  - On-demand online short courses provided by universities and non-governmental organizations
  - Live online short courses provided by private sector organizations
  - Other sustainability learning resources
- Management-themed learning resources
  - Project management
  - Root cause analysis
  - ISO management systems
- Online sustainability libraries and other information resources
  - General information
  - Green living
  - Environmental and social responsibility
  - Built environment, energy and technology, and
  - Sustainability in business and government.

1.10.4   The Book's Glossary

Like most specialty fields, sustainability is fraught with unfamiliar management and technical terms, buzzwords, and acronyms. The book's extensive glossary was prepared with less experienced business and technical readers in mind to provide them with readily available sustainability and management information.

1.10.5   One Size Does Not Fit All

The idea that one size does not fit all cannot be overstressed. Readers and organizations must adapt the book's prescriptions to suit their needs and circumstances.

For example, some will selectively apply prescriptions from discrete chapters to improve existing programs. Such improvements might include adding an annual sustainability program evaluation process, while others might involve an initiative to formally link strategic intentions to activities in each unit, function, and department. However, readers creating new comprehensive sustainability programs must delve deeply into the entire book. Regardless of how it is used, the book's prescriptions must be tailored to an organization's particular needs.

1.10.6 Different Readers Need Different Information

Although readers will benefit from a complete study of the book, time constraints may prevent it. Consequently, it is helpful to know the locations of information most relevant to specific roles in the initiation, design, implementation, and management of an SMS. To that end, Figure 1.2 suggests chapters and appendices that can be useful to specific categories of readers. The left column lists various types of SMS leaders, participants, and other stakeholders. Extending to the right are the preface, chapters, and appendices. The intersection marks on the matrix indicate essential information for that particular leader, participant, or stakeholder group.

1.10.7 Take One Chapter at a Time

Be forewarned that complete and even piecemeal read-throughs can be overwhelming, as can subsequent re-reads. The best approach is to become generally familiar with the contents but not attempt complete mastery from the outset. Consistent with adult learning theories, mastery will come when the step-by-step prescriptions are applied during the program creation's design, implementation, and management phases. So, avoid information overload and focus on one chapter at a time and only one task at a time within that chapter. Once each task is complete, then – *and only then* – move on to the next. Patience is a virtue when creating sustainability programs.

1.11 Chapter Takeaways

- Because its activities collectively and directly cause the most consequential environmental and social impacts – good and bad – the private sector must, out of necessity, lead the transition to a more sustainable economy. The private sector has the resources, expertise, and motivation at micro- and macro-economic scales required for rapid transformational change via the profit motive.

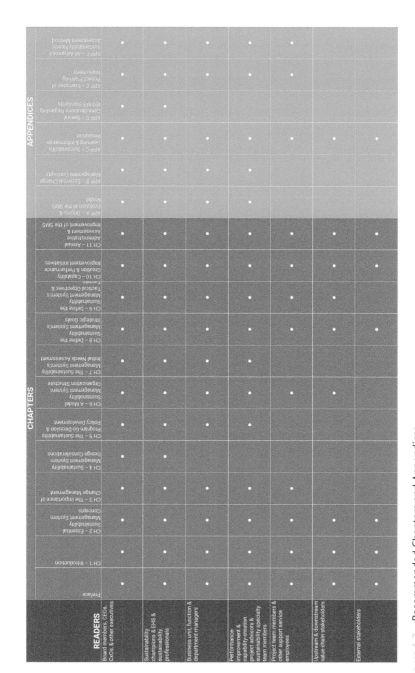

Figure 1.2 Recommended Chapters and Appendices.
Source: W. Borges.

- Sustainability is not a burden but a benefit when treated as an advanced approach to overall performance improvement. As such, it provides organizations with new management insights and methods for adding and creating value throughout the life cycles of their product and service value chains.
- Sustainability has the potential to improve all aspects of organizational performance throughout a value chain in the management concerns of:
  - Reduced risks
  - Improved efficiencies by eradicating wastes of all kinds
  - Increased revenues through product and service innovation, and
  - Enhanced tangible and intangible competitive advantages through transparency.
- At its most elemental level, this book is an OD textbook written to help change agents in their sustainability transformation efforts, whatever their title and position.
- The book provides organizations of various types and sizes with a practical stepwise design, implementation, and management model for a CI SMS to achieve sustainability goals with the least cost, effort, and risk. The model emphasizes:
  - Program evolution
  - Change management
  - Adaptability, and
  - Integration.
- The big questions for leaders assuming responsibility for creating a sustainability program are:
  - *How do you do it?*
  - *How much will it cost?* And,
  - *How long will it take?*
- The methods in this book prescribe ways to systematically and systemically:
  - Set an unambiguous sustainability policy
  - Define, prioritize, and shortlist an organization's most pressing sustainability needs
  - Meet those shortlisted needs within the organization's resource limits and opportunities with the least cost, effort, and risk, and
  - Produce significant, measurable benefits that exceed costs.
- Unlike systematic and systemic approaches to sustainability, common *ready-FIRE-aim* efforts risk failing to identify and meet an organization's most pressing needs, wasting scarce financial, labor, material, temporal, and tangible and intangible marketplace goodwill resources.

- The best management practices in SMSs include:
  - Policy-initiated mandates implemented through needs-assessed sustainability strategies, tactics, and initiatives cascaded throughout organizations using work processes based on the PDCA cycle, and
  - High degrees of proactivity, collaboration, and transparency.
- SMSs are major strategic initiatives requiring leadership determination and highly focused CM planning and actions. They take time to achieve the intended results; there are no shortcuts.
- It is possible to produce significant, measurable results early in an SMS's first year of operation. Early quick-win results to resolve important sustainability issues gain support for the SMS effort.
- None of the current international MS standards, ESG performance criteria, or regulations can provide an organization with a ready-to-implement SMS design. This is because they are not organizational design models.
- However, despite their shortcomings, standards, criteria, and regulations offer important SMS design considerations. Their underlying planning, organizing, controlling, and leading concepts are essential in designing and implementing an effective sustainability program that can fully integrate into an organization's overarching MS.
- Greenwashing is a rampant problem that damages reputations, risks regulatory and judicial sanctions, and ultimately threatens financial performance for organizations that knowingly or unwittingly employ the practice.
- The consequences for organizations that ignore the book's prescriptions can be severe. Their marketplace adversaries will exploit their inaction, losing competitive advantages, key talent, customers, stakeholder goodwill, and positive financial performance. These losses may not be immediate, but they are inevitable.
- Individuals and entire organizations can become discouraged by the effort and time required to design, implement, and manage an effective sustainability program. However, the book's advice can help newcomers start positively, and underperforming organizations finally achieve programmatic traction.
- Appendix C, *Sustainability Learning and Information Resources*, provides lists of sustainability-themed Masters of Business Administration degree programs, professional certificate and accreditation programs, specialty short courses, and other resources to help readers broaden and deepen their knowledge and skill sets in organizational management and sustainability.
- Although readers will benefit from a complete book study, time constraints may prevent it. Figure 1.2 suggests chapters and appendices that can be useful to specific categories of readers.

1.12   Further Reading

Supporting the book's ideas, the following seven discussions examine critical aspects of sustainability programs:

- Corporate governance and greenwashing gone awry
- Guidance on avoiding greenwashing
- How financially successful companies integrate ESG priorities to outperform their peers
- A CEO's guide to sustainability
- Ways to create business value with embedded sustainability
- Distorted views of sustainability that some business leaders hold, and
- Ways a major defense contractor benefited from its CI environmental MS.

1.12.1   The Volkswagen Emissions Scandal

Volkswagen Group's *Dieselgate* scandal is an example of corporate governance and greenwashing gone awry. Research in 2014 by the University of West Virginia's Center for Alternative Fuels Engines and Emissions uncovered one of the most significant scandals in automotive history. It revealed that the nitrogen oxide (NOx) emissions of several VW- and Audi-brand diesel cars were 40 times the EPA-permitted amount while driving on highways in the United States. In a deliberate act of deception, the company launched a long-term greenwash-based advertising campaign touting its *clean diesel* VW and Audi cars. The campaign was directed at environmentally conscious buyers who, at the time, viewed Volkswagen as an automotive leader in environmental stewardship. Corporate leadership ignored recommendations from engineers to install NOx controls. Instead, they mandated the installation of lower-cost monitor-defeating software to conceal high emissions during government tests.

The fallout from the highest corporate leaders' behavior included the following.

- At least 22 countries actively investigated Volkswagen.
- Total sanctions to date are estimated at $34.8 billion for fines, refits, and legal costs.
- Consumer protection penalties included buy-back, retrofit, repair, and other compensations for buyers of nearly 11 million vehicles worldwide.
- There have been prosecutions and convictions of some executives.
- An immediate decline in new car sales and a substantial decrease in its stock value, surprisingly affecting other German automaker stock values. The day after news of the scandal broke, Volkswagen's stock price declined by ~20%,

followed by another 12% drop the next day. A year later, Volkswagen's stock was down by 30%. In 2023, stock values remained almost half of the pre-scandal levels.

- Incalculable reputational damage reflected by ongoing financial difficulties and a new highly adversarial relationship with regulators.

Sources:

- www.statista.com/statistics/466109/annual-closing-share-prices-of -volkswagen/
- qz.com/volkswagen-dieselgate-scandal-rupert-stadler-audi-1850581201
- www.reuters.com/article/idUSKBN2141JA/
- www.justice.gov/opa/pr/volkswagen-spend-147-billion-settle-allegations- cheating-emissions-tests-and-deceiving
- knowledge.wharton.upenn.edu/podcast/knowledge-at-wharton- podcast/volkswagen-diesel-scandal/

Retrieved: 27 March 2024

1.12.2 Guidance: Making Environmental Claims on Goods and Services

The Government of the United Kingdom provides extensive guidance to help businesses comply with consumer protection laws when making environmental claims and thereby avoid greenwashing. Although the guidance is focused on environmental claims, it notes that it can also apply to the broader category of sustainability claims.

The essential principles are:

- Claims must be truthful and accurate
- Claims must be unambiguous
- Claims must not omit or hide important relevant information
- Comparisons must be fair and meaningful
- Claims must consider the entire life cycle of the product or service, and
- Claims must be substantiated.

In practical terms, when making a green claim, a business should be able to answer 'yes' or agree to each of the following statements:

- The claim is accurate and clear for all to understand
- There is up-to-date, credible evidence to show that the green claim is true
- The claim tells the whole story of a product or service or relates to one part of the product or service without misleading people about the other parts of the overall impact on the environment

- The claim does not contain partially correct or incorrect aspects or conditions that apply
- Where general claims (*eco-friendly*, *green*, or *sustainable*, for example) are being made, the claim reflects the whole life cycle of the brand, product, business, or service and is justified by the evidence
- If conditions (or caveats) apply to the claim, they're clearly set out and can be understood by all
- The claim will not mislead customers or other suppliers
- The claim does not exaggerate its positive environmental impact or contain anything untrue – whether clearly stated or implied
- Durability or disposability information is clearly explained and labeled
- The claim does not miss out or hide information about the environmental impact that people need to make informed choices
- Information that really cannot fit into the claim can be easily accessed by customers in another way (QR code, website, etc.)
- Features or benefits that are necessary standard features or legal requirements of that product or service type are not claimed as environmental benefits, and
- If a comparison is being used, the basis of it is fair and accurate and is clear for all to understand.

Source: Competition and Markets Authority, Government of the United Kingdom, 20 September 2021

www.gov.uk/government/publications/green-claims-code-making-environmental-claims/environmental-claims-on-goods-and-services
Retrieved: 25 March 2024

1.12.3 The Triple Play: Growth, Profit, and Sustainability

After examining the financial and ESG ratings of 2269 public companies, McKinsey's analysis found that financially successful companies that integrate ESG priorities into their growth strategies outperform their peers. However, the prerequisite is that they must also outperform on business fundamentals. It is important to note that not all industries achieved shareholder returns correlated with ESG-rating improvements.

The outperforming growth-profit-ESG companies tend to be guided by five principles, most of which echo the themes and concepts in this book:

- Integrating growth, profitability, and ESG into the core strategy
- Innovating ESG offerings to drive value creation
- Using mergers and acquisitions to capture ESG growth pockets rapidly
- Tracking and reporting ESG and related data transparently, and
- Embedding strategic priorities in the organizational DNA.

The analysts summarized their findings by saying:

Not only can you do well while doing good, but you can also do better.

Source: Doherty, R., Kampel, C., Koivuniemi A., et al, 9 August 2023, McKinsey & Company
www.mckinsey.com/capabilities/strategy-and-corporate-finance/our-insights/
the-triple-play-growth-profit-and-sustainability
Retrieved: 25 March 2024

1.12.4 The Visionary CEO's Guide to Sustainability

Bain's report states that executives know sustainability is a complex balancing act despite being presented with many simplistic answers. Employees and communities expect change, and companies are expected to take on their share of environmental and social challenges. The report reminds readers that executives are people, too, and many see this as their legacy.

Executives are concerned about the growing gap between their public commitments and their delivery on them. Some 75% of those surveyed believe they have not effectively embedded sustainability into their business. Bain also found that fewer than 40% of major companies across sectors are tracking their sustainability goals, including those related to water use, waste reduction, and preservation of biodiversity.

Top-of-mind goals for corporate boards and top management – and inspirational to employees – are proving difficult for organizations' profit and loss (P&L) owners – e.g. mid-level managers – who must reconcile immediate profit delivery with environmental and social commitments. They resist simplistic assurances that sustainability presents opportunities for enhanced performance, resulting in necessary trade-offs between vision and pragmatism.

The report offers valuable tools and viewpoints summarized by the following themes.

- During strategy development, focus on these three critical purpose, externalities, and shortages questions:
  - *What good do we bring to the world, and what is our purpose as a company?*
  - *What cost will humanity have to pay for us to grow?*
  - *What will get in our way, and what will we run short of?*
- Push for an "and" agenda focused on policy, technology, and behaviors.
- Listen to P&L owners and work to translate their struggles into team-sized challenges.

Source: Faelli, F., Lichtenau, T., Blasberg, J., et al., 25 October 2023, Bain & Company
www.bain.com/insights/topics/ceo-sustainability-guide/
Retrieved: 29 March 2024

1.12.5   Beyond Checking the Box: How to Create Business Value with Embedded Sustainability

A survey of 5000 executives across 22 industries in 22 countries during the second half of 2023 illustrates the folly of a "ready-FIRE-aim" approach to sustainability.

The survey found that spending on sustainability reporting exceeds spending on sustainability innovation by 43%, indicating that far too many organizations are approaching sustainability as a burdensome accounting or reporting exercise. In doing so, they fail to recognize this book's proposition that sustainability provides value chain transformation opportunities to:

- Improve risk management
- Reduce costs and expenses, especially those associated with energy consumption and waste generation
- Increase revenues through product and service innovation, and
- Gain tangible and intangible advantages through transparency.

The survey results suggest many organizations use misguided *cart-before-the-horse* approaches. The research also found that companies operationalizing sustainability throughout their organizations show better environmental, social, and financial outcomes. They are 52% more likely to outperform their peers on profitability and have a 16% higher revenue growth rate.

Source: Abbosh, O., Shim, C., Goos, E., et al, 29 February 2024, IBM Institute for Business Value, Research Insights
www.ibm.com/thought-leadership/institute-business-value/en-us/report/sustainability-business-value
Retrieved: 27 March 2024

1.12.6   The Burden of Proof for Corporate Sustainability is Too High

This article addresses distortions in some business leaders' views of sustainability, which lead them to blame sustainability for poor performance. Without understanding its business benefits, they erroneously believe that sustainability is a trade-off incompatible with financial responsibility that adds costs and drags down earnings. The article supports the book's methods in two important ways:

- Sustainability can cut expenses, slash risks, drive innovation and revenues, and build intangible brand value, and
- Sustainability should be embedded into organizational structures and processes.

Source: Winston, A., 3 October 2023, MIT Sloan Management Review
sloanreview.mit.edu/article/the-burden-of-proof-for-corporate-sustainability-is-too-high/
Retrieved: 27 March 2024

1.12.7 General Dynamics Zero Waste Program

General Dynamics' industry-leading Zero Waste program reduced hazardous and other wastes by over 75% in just four years. One division achieved over a 90% reduction. The efficiency improvements produced significant cost savings and improved the company's reputation with regulators, stakeholders, and the general public. Some unexpected benefits included preferential treatment in contract awards, reduced regulatory risks and sanctions, positive press, and numerous recognitions and awards for environmental stewardship, quality, and business performance excellence.

Source: J. Grosskopf, 2024

2

Essential Sustainability Management System Concepts

This chapter introduces these essential *sustainability management system* (SMS) concepts:

- New organizational structures are needed to drive the evolution of an effective sustainability culture
- Based on *systems thinking* and *rational decision-making*, SMSs concentrate on continuous incremental and breakthrough organizational performance improvement by addressing 3P concerns throughout their value chains, and
- An effective SMS *hardwires* strategic intentions to each of an organization's operational and administrative activities.

Figure 1.1 concept map introduced a highly generalized view of the SMS model. This chapter concludes by expanding the SMS model with a detailed process map illustrating its stepwise functions and crucial decision points. This thorough process map provides the graphic basis for the SMS design prescriptions in subsequent chapters.

2.1 The Relationships Between Organizational Structure, Performance, and Culture

Anyone who has participated in a performance improvement initiative is likely familiar with this concept promoted by such luminaries as Drs. W. Edwards Deming, Donald Berwick, Paul Batalden, and others:

> *Every system is perfectly designed to get the results it gets.*

Sustainability Programs: A Design Guide to Achieving Financial, Social, and Environmental Performance, First Edition. William Borges and John Grosskopf.

Therefore, the system's design must be changed to achieve different results. Underneath this simple proclamation are these essential management concepts:

- A well-designed and managed organizational structure actively drives desirable behaviors
- Conversely, a poorly designed and managed organizational structure enables undesirable behaviors leading to underperformance or failure, and
- Collective behaviors over time determine – for better or worse – performance quality in the short term and organizational culture in the long term.

Figure 2.1 shows some of the more important elements of organizational structure where changes can be made to produce measurable results in any *management system* (MS).

Successful SMS development involves determined changes via planning, organizing, controlling, and leading to achieve the organization's sustainability goals with the least effort, cost, and risk. These four activities are bedrock management functions.

Figure 2.1 Typical Organizational Structure Elements.
Source: W. Borges.

2.2 Key Concepts

Before proceeding, it is necessary to define sustainability and several of its underlying concepts in organizational management terms. The book's glossary provides additional detail on many of these concepts. Chapter 1 provided this organizational management definition of sustainability:

> *Throughout the entire closed-loop lifecycle of a product or service, sustainability is the way an organization creates value by maximizing the positive social, environmental, and economic effects of its activities while minimizing their adverse impacts.*

Further, as classic examples of systems thinking, CI MSs clearly illustrate rational decision-making.

> *Rational decision-making is the opposite of intuitive decision-making. It is a strict procedure using objective knowledge and logic. It involves identifying the problem, gathering facts, analyzing them, identifying options, and considering their ramifications before deciding.*

In general, an MS is:

> *A collection of interrelated, interdependent, and interacting functions to establish policies, processes, and activities to achieve goals that respond to an organization's needs.*

Specifically, a CI MS is:

> *Structured on a planned cyclical schedule to produce improvements to processes, products, and services through incremental and breakthrough performance improvements.*

Building on this concept, the book's SMS model prescribes rapid-cycle processes and schedules to produce incremental and breakthrough 3P performance improvements throughout an organization's value chain. As illustrated in Figure 2.2, an SMS focuses on:

- Enhancing a system's administrative effectiveness
- Managing risks, including compliance with global, regulatory, contractual, and industry standards for financial performance, product and service quality, social responsibility, and environmental protection

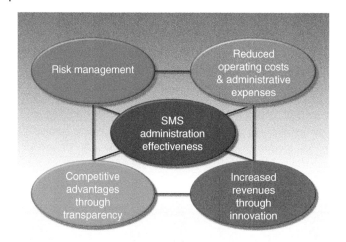

Figure 2.2 Business Focus in an SMS.
Source: W. Borges.

- Reducing costs and expenses, especially those associated with energy consumption and waste generation
- Growing revenues with innovative environmental and social-attribute products and services, and
- Building tangible and intangible competitive advantages through transparency.

In the relatively brief history of sustainability as a distinct management discipline, environmental and financial factors have been typically emphasized more than social aspects. However, social factors have not been ignored. They have been considered separately as *corporate social responsibilities* (CSR). CSR is a broad concept that promotes positive actions regarding ethics, justice, diversity, equality, and general well-being for all stakeholders involved internally in an organization's value chain, as well as those affected externally by it. In current practice, though, CSR is combined with environmental and financial concerns in a unified sustainability discipline.

2.3 Management Perspectives on Building an Effective Sustainability Program

If you talk to one expert, sustainability is all about facility design. However, when you speak with others, you will learn it is about energy and *greenhouse gases* (GHG), water resources, waste *reduction, reuse,* and *recycling* (3Rs), green information technology, sustainable supply chain management, or greenwash-free marketing and public relations.

Who is right? Collectively, all of them. Individually, none of them. So, beware when listening to experts … *and this caution extends even to this book's authors.* Technical specialists tend to view broad disciplines, such as sustainability, in terms of their own narrowly defined professional interests. There is nothing wrong with that. Specialists are expected to have this kind of laser-like focus.

However, when an organization starts working on sustainability, it needs to take a *SWOT* (pun intended) to determine all its *strengths, weaknesses, opportunities,* and *threats.* (See Chapter 7) It cannot do that well if it distracts itself by prematurely focusing on only a few of the discipline's specialty topics. The risks of a narrow focus or mistargeted effort are missed opportunities, ineffectual results, and wasted resources, which end in frustration with program underperformance or, even worse, failure.

An organization's SMS needs to consider dozens of specialties and issues. The major categories of specialties and concerns are shown in Figure 2.3, which only hints at the number of concerns an SMS might need to address. Details are listed later in Figures 7.1, 7.2, 7.3, and 7.4.

With dozens of sustainability specialties and issues, how does an organization decide which ones to work on? It is done through a CI SMS that links – some say

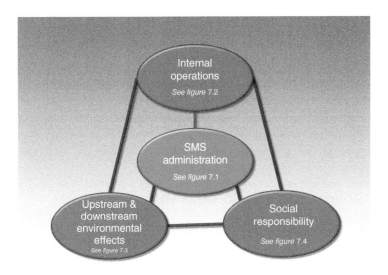

Figure 2.3   Typical Concerns in an SMS.
Source: W. Borges.

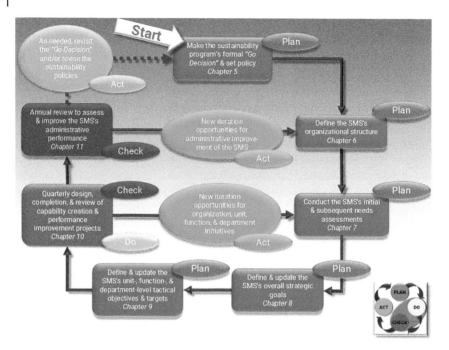

Figure 2.4   Detailed Process Map: The SMS Model.
Source: W. Borges.

hardwires – its strategic intentions to each of its operational and administrative activities. Like the one shown in Figure 2.4, it is an MS that:

- Defines, prioritizes, and shortlists its most pressing sustainability needs, ideally every fiscal quarter
- Assigns specific project and other initiative accountabilities
- Designs adequately resourced projects to meet its highest-priority needs
- Actively monitors project progress and takes immediate and effective corrective actions when variances and non-conformities occur, and
- Celebrates and rewards successes and applies lessons learned from those successes – as well as any failures – to future efforts.

Elaborating on Figure 1.1 concept map, Figure 2.4's detailed process map previews the subsequent chapters' SMS development and administration steps. Notice that the activities are labeled to show how the *plan-do-check-act* (PDCA) elements are related to the individual steps.

2.4   Chapter Takeaways

- Every system is perfectly designed to get the results it gets. So, to achieve different results, the current organizational system's design must be changed.
- Successful SMS development involves systematic planning, organizing, controlling, and leading to achieve the organization's sustainability objectives with the least effort, cost, and risk.
- At the very least, an SMS is focused on improving organizational performance through its administrative effectiveness, risk management, cost and expense reduction, revenue growth through innovation, and competitive advantages through transparency.
- The SMS's focus is consistent with the 3P concept, wherein environmental and social concerns are integrated and considered along with financial ones.
- When starting work on sustainability concerns, an organization must determine its strengths, weaknesses, opportunities, and threats. A poorly conducted needs assessment risks missed opportunities, ineffectual efforts, and wasted resources, ending in frustration with underperformance or program failure.
- An effective SMS hardwires its strategic intentions to each operational and administrative activity by:
  - Defining, prioritizing, and shortlisting the organization's most pressing sustainability needs
  - Assigning accountabilities for specific projects and other initiatives
  - Designing adequately resourced projects to meet the highest-priority needs
  - Actively monitoring project and other initiative progress and taking immediate and effective corrective actions when variances and non-conformities occur, and
  - Celebrating and rewarding successes and applying lessons learned from those successes – and failures – to future efforts.

2.5   Further Reading

Supporting the book's ideas, the following three discussions examine characteristics of successful sustainability programs:

- Organizational traits of the organizations benefiting from environmental, social, and governance (ESG);
- Characteristics of the world's best corporate sustainability programs; and
- Attributes of sustainable business models.

This McKinsey survey of more than 1,100 respondents in 90 countries revealed that nine in ten respondents said ESG subjects are on their organizations' agendas. Respondents also reported their organizations have created financial value and increased broader impact from ESG, two conditions McKinsey calls "ESG momentum." The survey report describes seven organizational traits of the organizations benefiting from ESG subjects that echo this book's themes, concepts, and methods. The organizations:

- Approach ESG from a business growth perspective and focus on identifying and exploiting new opportunities
- Work to connect with external stakeholders and be accountable to them
- Identify and prioritize specific stakeholder priorities as core parts of the business strategies
- Empower a specific executive in the C-suite to work with the CEO in defining and achieving ESG aspirations
- Build a central, but not necessarily large, ESG team that brings together talent from across the organization to help meet ESG goals
- Embed ESG purposes into multiple aspects of their business, and
- Tie ESG metrics to compensation, using key performance indicators to gauge progress on objectives.

Source: Korkmaz, B., Nuttall, R., Pérez, L., et al, 26 May 2023, McKinsey & Company

www.mckinsey.com/capabilities/strategy-and-corporate-finance/our-insights/esg-momentum-seven-reported-traits-that-set-organizations-apart

Retrieved: 28 March 2024

Consistent with this book's prescriptions, this article lists a set of sustainability program characteristics shared by early adopters and describes how they view themselves as corporate citizens and interact with society.

- Corporate responsibility and environmental responsibilities are embedded in company mission statements and charters.
- Companies recognize their long-term societal and environmental impacts and incorporate them into their strategic planning and risk management processes.

- Senior management teams and boards of directors actively guide companies' CSR programs, goal setting, and sustainability reporting.
- Companies aspire to be the best in their industries, internally and externally, in governance, sustainable product development, ethical business practices, and environmental stewardship.
- Companies transparently report their sustainability performance, both successes and shortcomings.
- The connection between companies' long-term prosperity and CSR is effectively articulated.
- CSR champions are consulted and rewarded.
- Stakeholder engagement is authentic and leads to change.
- CSR performance metrics follow international standards.
- Companies take a long-term view.

Source: McIntire, L., 11 March 2015, CSRWire Blogs, Triple Pundit Newsletters

www.triplepundit.com/story/2015/ten-characteristics-worlds-best-corporate-sustainability-programs/87481

Retrieved: 22 May 2024

2.5.3 How to Tell if Your Business Model is Truly Sustainable

As a part of an ongoing Boston Consulting Group series that describes the concept of *sustainable business model innovation*, this article presents the findings from a research project involving more than 100 business models focused on business value and environmental and social benefits. The research findings were summarized as nine attributes of sustainable business models. The first six attributes can apply to many business models independently of environmental or social impact, at least until issues arise from new regulatory constraints or stakeholder backlash. The last three attributes provide the core of sustainable business models. Consistent with the book's emphasis on organizational performance, a sustainable business model:

- Scales effectively without increasing risks or diminishing returns
- Increases differentiation and competitiveness
- Reduces the potential for commoditization
- Uses network effects to achieve growth and multiply the value
- Harnesses business ecosystems for advantage
- Remains durable against environmental and societal trends
- Creates environmental and societal benefits material for key stakeholders

- Increases returns to shareholders and environmental and societal benefits to stakeholders, and
- Animates purpose.

Source: Young, D., and Gerard, M., 9 April 2021, BCG Henderson Institute, Boston Consulting Group,
www.bcg.com/publications/2021/nine-attributes-to-a-sustainable-business-model
Retrieved: 22 May 2024

3

The Importance of Change Management

As stated throughout this book, SMSs are strategic initiatives that cascade through an organization into all units, functions, and departments. However, if an organization fails to adequately prepare for and institutionalize change at all levels, the entire initiative is doomed to underperform or fail. Despite this reality, formal CM is often an underappreciated and overlooked element of SMS design, implementation, and management. CM is:

> *A systematic approach to planning, initiating, realizing, controlling, stabilizing, and sustaining new and altered work activities at the organizational, group, and personal levels.*

3.1 Structure, Behavior, Performance, and Culture

The book's *engineering approach* focuses on management structure factors such as system models, strategies, processes, and job roles. Indeed, new structural elements are essential in avoiding or mitigating significant obstructions to organizational change that are illustrated by Dr. Peter Drucker's adage:

> *Culture eats strategy for breakfast.*

Professor Drucker did not suggest that organizational structures such as strategies are unimportant. Instead, he warned ominously that an organization's culture has a strong potential to obstruct changes required to implement strategies. While the adage does not suggest it directly, the converse is also true, i.e.:

> *Cultures can accommodate and even drive change when organizations are appropriately structured and managed.*

Sustainability Programs: A Design Guide to Achieving Financial, Social, and Environmental Performance, First Edition. William Borges and John Grosskopf.

That is why this book stresses the maxim attributed to Drs. W. Edwards Deming, Donald Berwick, Paul Batalden, and others:

Every system is perfectly designed to get the results it gets.

In keeping with this reality, a management system must be changed to achieve different results, especially when dealing with change-resistant culture issues. For

Change is good, you go first.

Scott Adams' Dilbert

that reason, it is essential to remember the dynamic relationships between *organizational structure, behaviors, performance quality,* and *culture* introduced in Chapter 2:

Organizational structure – or the lack thereof – actively drives or passively enables good and bad behaviors.

Collective behaviors over time determine – for better or worse – performance quality in the short term and organizational culture in the long term.

However, the structured *engineering approach* is only part of the CM effort to create an effective SMS. The other equally important factor is the *psychology approach*, where well-planned and executed CM efforts are focused on organizational, group, and individual behaviors.

3.2 A Change Management Appendix

Although this book does not provide detailed CM prescriptions, it is critically important to introduce some of the discipline's fundamental ideas. Appendix B, *Essential Change Management Concepts*, provides CM concepts, recommendations, warnings, and encouragements. Although the appendix is lengthy, it barely scratches the surface of the highly nuanced CM and organizational development disciplines.

For some readers, Appendix B is an essential CM primer; it is a refresher course for others. In either case, it summarizes many ideas an organization needs to employ to successfully design, implement, and manage an effective continuous improvement SMS. Readers should also explore the extensive body of available CM literature along with the learning resources listed in Appendix C.

The essential organizational development and CM concepts presented in the appendix are:

- Overview of change management
- Change management success and failure factors

- Schools of thought within change management
- Essential change management concepts
- The role of a change agent
- Preparations for change
  - A basic change process
  - Communications planning
  - Change readiness assessments
  - Readiness assessment data acquisition
  - Anticipating and managing obstructions
    - The root cause of change resistance
    - Typical reasons for change resistance
    - Types of change resistance
    - Managing the TANSTAAFL dilemma
    - Leadership shortcomings that lead to obstructionist behaviors
    - Methods for overcoming resistance in preferred order
- Institutionalizing change, and
- Closing thoughts on change management.

I'm in favor of progress; it's change I don't like.

Samuel Clemens

3.3 Chapter Takeaways

- Change management is a systematic approach to planning, initiating, realizing, controlling, stabilizing, and sustaining new and altered work activities at the organizational, group, and personal levels.
- However, CM is often an underappreciated and overlooked element of SMS design, implementation, and management.
- If an organization fails to adequately prepare for and institutionalize change at all levels, the entire initiative is doomed to underperform or fail.
- This book stresses the maxim attributed to Drs. W. Edwards Deming, Donald Berwick, Paul Batalden and others: "Every system is perfectly designed to get the results it gets."
- Organizational structure – or the lack thereof – actively drives or passively enables good and bad behaviors.
- Collective behaviors over time determine – for better or worse – performance quality in the short term and organizational culture in the long term.
- New organizational structure elements are essential in avoiding or mitigating significant obstructions, like those illustrated by Dr. Peter Drucker's adage, "Culture eats strategy for breakfast."

- The converse is also true, i.e. cultures can accommodate and even drive change when organizations are appropriately structured and managed.
- An engineering approach is only part of the CM effort to create an effective SMS. The other equally important approach, the psychology one, is a focus on organizational, group, and individual behaviors.

3.4 Further Reading

Supporting the book's ideas on change management, the following four examples discuss critical aspects of change management:

- Why the speed of change is crucial to transformation
- Why structured change management is essential to sustainability program success
- A summary of key change management concepts for sustainability program leaders, and
- The need to create a change-ready organization by developing an informed and skilled workforce.

3.4.1 Ready, Set, Go, and Keep Going: Why Speed is Key to a Successful Transformation

This article from McKinsey argues that companies seeking to transform all or part of their businesses in rapidly changing environments must create value quickly.
Source: Greco, L., and Silverman, Z., 4 May 2023, McKinsey & Company
www.mckinsey.com/capabilities/transformation/our-insights/ready-set-go-and-keep-going-why-speed-is-key-for-a-successful-transformation
Retrieved: 6 June 2024

3.4.2 Why 'Change Management' is Essential for Your Sustainable Ambitions: Three Pillars: From Sustainability Ambitions to Sustainable Results

This PwC blog post reinforces the book's assertion that change management is essential to bridging the gap between sustainability strategy and measurable performance. Further, it stresses that a structured change management approach is as much about sustainability as it is about transformation.

The authors suggest that sustainability change management efforts should include these three individual-centric ideas:

- *Engage Me*: Does leadership allow me to prioritize sustainability in my day-to-day work?

- *Equip Me*: What knowledge, skills, and tools do I need?
- *Engage Me*: What information do I need to know, and what is the desired mindset?

The foundational change management ideas discussed in the post are:

- Start with leadership to drive the sustainability strategy
- Leverage the organization's network of sustainability enthusiasts
- Break transformation activities into bite-size chunks, and
- Create partnerships to work toward shared goals.

The post concludes with one of this book's admonishments:

> *You are not seeking perfection but progress. Don't let the desire for perfection get in the way of starting and learning by doing in your journey towards a more sustainable business.*

Source: Isfordink, P., and Tan, V., 17 July 2023, PWC Netherlands
www.pwc.nl/en/topics/blogs/why-change-management-is-essential-for-your-sustainable-ambitions.html
Retrieved: 1 April 2024

3.4.3 Change Management for Sustainability

In addition to amplifying many of the book's SMS concepts, the blog post from Prosci Inc., a leading CM consulting firm, offers eight high-level tips:

- Walk the talk of sustainability and be an effective sponsor
- Communicate the "why" of sustainability and specific projects clearly
- Harness sustainability-oriented innovation, start small, and think long-term
- Empower grassroots ideas through ideation campaigns
- Make it a two-way conversation and understand where the impacts are
- Show progress and celebrate successes
- Support people through change, and
- Follow a structured and intentional approach to change management.

Source: McNeive, A., 17 January 2024, PROSCI, Inc.
www.prosci.com/blog/change-management-for-sustainability
Retrieved: 1 April 2024

3.4.4 Deloitte Leads Way on Employee Upskilling in Sustainability

Deloitte Touche Tohmatsu recognizes that clients are increasingly seeking new advantages through practical sustainability, climate, and equity problem-solving.

Additionally, many Gen Z and Millennials expect their employers to empower them with training and support for making sustainable decisions in their personal lives and developing the skills required to transition to a low-carbon economy.

Deloitte recognizes this expectation as it is one of the first big companies to roll out a worldwide climate learning program for all 457,000 employees. Specifically, the company is advancing and accelerating these skills among its more than 150,000 professionals in the United States. Deloitte's employees have responded to the organization's on-demand internal learning offerings focused on sustainability, climate, and equity knowledge by making it the number one bookmarked resource.

Regarding its client service offerings, Deloitte's performance improvement goal is integrating core advisory capabilities with innovative sustainability thinking to gain a competitive edge. The firm invested $1 billion into a sustainability and climate practice to expand and elevate its robust bench of learning and development capabilities in critical in-demand skills, such as sustainability literacy and green skills. It has expanded collaborations with renowned academic institutions to develop and roll out a differentiated, integrated learning program. These institutions include MIT Sloan School of Management, NYU Stern Executive Education, the Center for Sustainable Business, and Arizona State University.

Source: Birch, K., 13 February 2024, Sustainability Magazine
sustainabilitymag.com/sustainability/deloitte-leads-way-for-employee-upskilling-in-sustainability
Retrieved: 1 April 2024

4

Sustainability Management System Design Considerations

This chapter poses SMS design questions, the answers to which will help leaders create an effective sustainability program tailored to the unique needs of their organizations. Note that whether an organization uses all, some, or none of the book's prescriptive design steps presented in subsequent chapters, it will still need to answer the questions in this chapter at various times during sustainability program development. These questions prompt leaders to consider, customize, and apply essential, highly flexible concepts that many leading organizations use to become sustainable.

It is important to note that these SMS design questions use the same basic concepts underlying most *continuous improvement* (CI) *management system* (MS) standards discussed in the Preface, Chapter 5, and Appendix D.

Further, the design questions integrate conventional *organizational development* (OD) considerations because this is a transformational process. These OD considerations require the involvement of CM, *training and development* (T&D), *performance improvement*, and other specialists in the earliest stages of SMS development.

> *... an organization must have goals, take actions to achieve those goals, gather evidence of achievement, study and reflect on the data and from that take actions again. Thus, they are in a continuous feedback spiral toward continuous improvement ...*
>
> **W. Edwards Deming**

Initial answers to these design questions produce high-level ideas for fundamental SMS structures and processes. Subsequent chapters explore the SMS model's administration and operations design details.

Sustainability Programs: A Design Guide to Achieving Financial, Social, and Environmental Performance, First Edition. William Borges and John Grosskopf.
© 2025 John Wiley & Sons, Inc. Published 2025 by John Wiley & Sons, Inc.

4.1 Cautionary Notes Regarding SMS Design Flexibility

Refer back to this adage introduced in Chapter 2:

> *Every system is perfectly designed to get the results it gets.*

Because of this reality, the SMS model is prescriptively detailed to produce incremental and breakthrough sustainability performance via capability creation and performance improvement initiatives. Indeed, the model has significant flexibility. However, when taking advantage of that flexibility, there are also substantial risks to SMS effectiveness. Remember the corollary to this adage in Chapter 2:

> *A poorly designed and managed organizational structure enables undesirable behaviors leading to underperformance or failure.*

One of the purposes of the prescriptive detail is to limit wiggle room in the SMS's processes wherein personnel who – because of other demands on their time and attention – can procrastinate on their assigned accountabilities or ignore them altogether. Therefore, when exercising design flexibility, leaders should constantly consider how modifications to the SMS model as prescribed in subsequent chapters might result in less-than-desirable sustainability program outcomes. This is especially important when designing support processes to ensure personnel can successfully achieve time-constrained accountabilities.

4.2 A Format for Answering SMS Design Questions

As shown in Figure 4.1, the SMS design questions should be answered in detail using the conventional project management format of *5Ws & 1H + Check* – i.e. *what, why, who, where, when,* and *how,* plus *check*. Note that the *check* activity in the figure mentions *MDC Methods*, i.e. *monitor-detect-correct*. The *monitoring* and *detecting* activities are the *check* step of the *plan-do-check-act* (PDCA) *cycle*. The MDC method's *correct* activity is the PDCA cycle's *act* step. These are the MDC activities:

- *Monitoring* is both programmatic and in-the-moment observation of work in progress
- *Detecting* is the immediate response of documenting and reporting discovered variances and nonconformances, nonconformities, or noncompliances to plans, standards, or regulations, and
- *Correcting* is the immediate and effective corrective actions taken in response to plan, standard, or regulation variances and nonconformances, nonconformities, or noncompliances.

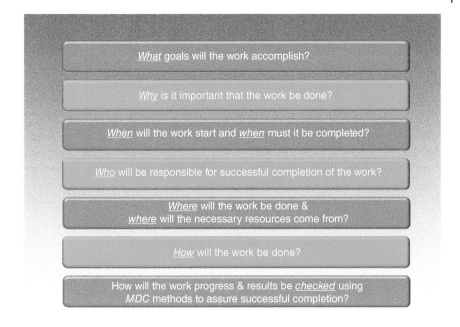

What goals will the work accomplish?

Why is it important that the work be done?

When will the work start and when must it be completed?

Who will be responsible for successful completion of the work?

Where will the work be done &
where will the necessary resources come from?

How will the work be done?

How will the work progress & results be checked using
MDC methods to assure successful completion?

Figure 4.1   The 5Ws & 1H + Check Format.
Source: W. Borges.

Distinctions Between Variances and Nonconformances, Nonconformities, and Noncompliances

Variances occur when a standard or plan is followed, yet the results of the work effort are not as expected. In contrast, nonconformances, nonconformities, and noncompliances happen when a plan, standard, or regulation is not followed, and the results of the work effort are different than expected. Although such results are typically negative, occasionally they are positive. When the results of variances and nonconformances and nonconformities are positive, they should be highlighted as lessons learned in the periodic project-progress reports and the end-of-project closure report. However, regulatory noncompliances with positive results raise risk management concerns. See Glossary for more complete discussions on the distinctions between *Variances* and *Nonconformances, Nonconformities, and Noncompliances.*

Including MDC methods in the SMS design avoids much procedural guess-work when new processes fail to go as planned. This is a critically important consideration in any new OD initiative. MDC methods provide immediate feedback to correct flawed processes and work performances that otherwise may go undetected or ignored, reducing an SMS's overall potential for success.

Concise answers to the SMS design questions in the *5Ws & 1H + Check* format will produce a customized SMS with the potential to be effective from the outset. Later, the initial SMS design will become more refined and even more effective because of the iterative CI methods on which it is based.

SMS activities are ideally performed during a 13-week fiscal quarter. This makes it possible to integrate SMS activities into an organization's enterprise-wide MS. Organizations with different timeframes in their planning and control cycles must adjust their SMS designs.

4.3 High-level System Plan Phase: SMS Initiation

Before an organization commits its limited resources to sustainability initiatives, it needs to decide systematically what it needs to focus on and how it will successfully achieve its goals. The questions in the following sections address important considerations an organization must make initially in designing, implementing, and managing an SMS.

4.3.1 Create the SMS and Assure Executive Buy-In

- *How will senior leadership – i.e. board of directors members and C-level officers in the private sector and elected and appointed officials in the public sector – initiate a sustainability program and set its governing policies?*
- *Who among the board of directors members, C-level officers, other executives, or public sector officials will be the SMS champion(s)?*
- *How will time be set aside for executives, officials, managers, specialists, and other participants to focus on designing, implementing, and managing the SMS?*
- *How will senior leadership ensure adequate budgetary, personnel, and other resources to implement the mandated policies, procedures, goals, objectives, targets, projects, and related initiatives?*
- *How will executive, managerial, and other participant accountabilities for SMS success be set and incentivized?*

4.3.2 Create the SMS's Shared Governance Function

The shared governance concept refers to structures and processes through which all organizational levels and ranks participate in the development, execution, and implementation of an SMS's policies, goals, objectives, and targets.

- *How will the SMS's organizational structures, including functions, lines of authority, participant roles, and work processes be designed?*
- *How will internal and external stakeholders be prepared to participate competently, willingly, and enthusiastically in the SMS?*

4.3.3 Create an Iterative Process to Define Near- and Long-term Sustainability Priorities

- *How will senior leadership identify, prioritize, and shortlist the organization's most pressing sustainability needs within its planning cycles?*
- *In future iterations of the SMS, how will lessons learned be documented, used to set new sustainability priorities, and improve the system itself?*

4.3.4 Create a Transparency Process to Monitor, Assess, and Report Sustainability Program Performance

- *How will policies, goals, objectives, targets, procedures, budgets, accountabilities, and SMS outcomes be made readily available on a need-to-know basis to the full range of internal and external stakeholders?*

4.3.5 Define the SMS Implementation Process

- *How will senior leadership – with the active participation of OD, CM, T&D, performance improvement, and other specialists – define and manage the SMS introduction process, including phases to:*
  - *Assess the organization's readiness for change*
  - *Prepare the organization to change*
  - *Start the SMS work processes and*
  - *Follow-up and reinforce SMS skill sets to achieve measurable leadership, unit, function, department, and personal competencies, thereby contributing to SMS success.*

4.4 Operations-level Plan Phase: SMS Administration

Once an SMS is designed, the organization's most pressing sustainability needs must be resolved through specific time-sensitive work activities completed by accountable leaders and staff. The following question addresses vital considerations in that process.

- *How will senior leadership – with the active participation of key internal and external stakeholders – define and cascade specific accountabilities for sustainability goals, objectives, and targets, including returns on investment and other performance requirements, down through the organization?*

4.5 Operations-level Do-and-Check Phase: Capability Creation and Performance Improvement

Note that capability creation and performance improvement initiatives are typically called *projects*. Specific projects resolve an organization's most pressing needs within an organization's administration and operations units, functions, and departments. Answers to the following questions will provide successful project designs and outcomes.

4.5.1 Create a Time-constrained Process to Design and Successfully Complete Specific Projects

- *How will single- and multiple-quarter sustainability capability creation and performance improvement projects be designed and completed at all unit, function, and department levels?*
- *How will sustainability capability creation and performance improvement project performance be monitored and reported to confirm that goals, objectives, and targets are on track to successful completion?*

4.5.2 Create Quarterly Processes to Evaluate Projects

- *How will projects lasting more than one quarter be evaluated?*
- *How will end-of-project reviews be conducted to evaluate the degree of project success?*
- *When projects fall short, how will corrective actions be defined and implemented?*
- *How will project successes be celebrated and rewarded?*

4.6 Operations and High-level Act Phase: SMS Enhancement

The objective evaluation of the MS itself is a key feature of any effective MS. These evaluations help discover additional opportunities for improvement to be implemented during the next SMS iteration. They improve the SMS and provide substantial information for *environmental, social, and governance* (ESG) evaluations and reporting.

4.6.1 Create Processes to Improve the SMS

- *How will lessons learned from projects – especially the definition of new best practices – be documented, communicated, and successfully implemented?*

- *How will the list of the organization's most pressing sustainability needs be updated and re-prioritized quarterly and annually?*
- *How will lessons learned regarding the efficiency of the SMS – i.e. the ability to achieve sustainability aspirations with the least cost, effort, and risk – be used to improve it?*
- *How will the next iteration of the PDCA-based SMS be initiated for the next fiscal quarter, fiscal year, or other period?*

4.7 Some Additional Thoughts on Change Management

An SMS's design, implementation, and management are significant OD efforts with enormous challenges. Successful responses to these challenges reside in the CM details. It is critical that an organization's SMS champion and other responsible leaders anticipate and continually address the concerns of stressed and often pessimistic employees who reflexively resist seemingly never-ending streams of change initiatives.

Routine organizational changes are complicated enough. People must stop doing what they are currently doing and start doing something new. They are expected to accept such changes even when they would rather not due to the demands of current responsibilities. Transforming an organization into a sustainable one is not routine; it is far from it. Because of the depth and breadth of changes required to introduce and run an SMS, it is a unique challenge for boards of directors, C-level officers, strategic business unit leaders, middle managers, and in-house OD, CM, T&D, and performance improvement specialists.

However, it is doable and *sustainable* in an organizational management sense. It requires a well-designed and effectively executed CM effort involving determined, proactive, hands-on, and steadfast leadership at all levels. CM concepts are also discussed contextually in Chapters 3 and 5, and elaborated on in Appendix B, *Essential Change Management Concepts.*

> **Coming together is a beginning; keeping together is progress; working together is success.**
>
> **Henry Ford**

4.8 Sustainability Software Applications and SMS Documentation Boilerplate

It is necessary to clarify a common misconception regarding SMSs and sustainability software applications called *systems*. Regardless of what a vendor may suggest, sustainability information management software applications are not SMSs. Software packages cannot possibly provide an organization with a ready-to-go SMS customized to its specific needs.

However, SMS software packages are helpful information management tools. The trick is knowing which ones are best suited to an organization's needs. It is easier for leaders to know which ones might be most valuable once they have managed the SMS through at least one fiscal quarter or other performance period cycle. Waiting to evaluate software needs is prudent in avoiding ineffectual information management efforts, unnecessary costs, wasted time, and frustration throughout an organization.

Indeed, judiciously selected software applications are helpful, cost-effective SMS tools. The best ones are light-years ahead of manual methods, spreadsheets, and other in-house information management tools. Of course, a key consideration in selecting any software package is its compatibility with the organization's *enterprise resource planning (ERP) system*. So, once the SMS's information management needs are defined, select carefully, spend wisely, implement methodically, and manage competently.

Thanks to advances in *big data analytics* (BDA) and generative *artificial intelligence* (AI), new software applications for sustainability are developing rapidly. The current state of that development is discussed in Section 4.10.

Vendors also offer ready-made boilerplate packages containing single-purpose MS policies, procedures, and processes intended to achieve a specific industry standard certification, such as ISO 14001 or ISO 45001. These documentation packages are neither MSs nor SMSs. Instead, as formatted collections of generic text, they are merely intended to help an organization comply with an industry standard's rudimentary outline and content requirements for certification. They cannot provide an organization with an implementable customized SMS to produce measurable people, planet, and profit results.

4.9 Chapter Takeaways

- This chapter presents the critical design questions that must be answered to create a customized SMS.
- The SMS design process uses the concepts underlying most formal CI MSs and integrates conventional OD considerations.
- The general SMS features developed by this chapter's question-based design process are detailed in later chapters.
- The design process uses the 5Ws & 1H + Check format. The *check* concept stresses the importance of monitoring, detecting, and correcting variances and nonconformances to performance standards.

- The design process produces an SMS operating ideally on a 13-week fiscal quarter basis within a fiscal year.
- The SMS design topics correspond to the PDCA cycle, with the process divided into four phases:
  - High-level system plan phase – SMS creation
  - Operations-level plan phase – SMS administration
  - Operations-level do-and-check phase – Capability creation and performance improvement, and
  - Operations- and high-level act phase – SMS enhancements.
- An SMS is a significant OD effort with enormous challenges.
- It is critical that an organization's leaders and its various OD, CM, T&D, and performance improvement specialists formally work proactively with all stakeholders to minimize resistance to the SMS.
- Contrary to what some vendors may suggest, sustainability software applications are not SMSs.
- Software packages are helpful SMS information management tools. However, it is prudent to delay any selection until the organization has completed at least one fiscal-quarter cycle.
- Once SMS information management needs have been defined, carefully select, spend wisely, implement carefully, and manage competently.

4.10 Further Reading

Supporting the book's ideas are three discussions on SMS design:

- Steps to consider when creating a sustainability program
- The importance of sustainability metrics, and
- The current state of advanced sustainability information technology.

4.10.1 12 Steps to Start Your Corporate Sustainability Program: A Guide to Help You Get Your Company's Sustainability Program off the Ground

Sustain.Life, a software and advisory services firm, suggests 12 steps to consider when starting a corporate sustainability program. These developmental steps closely align with this book's SMS model.

- Understand what's important to the company's stakeholders. Establish the right priorities through a structured, formal stakeholder engagement process.
- Identify key sustainability areas and set priorities with a materiality matrix. Addressing every sustainability issue is impossible, and trying to do so would result in a lack of program focus and the potential for meaningful change.

- Create a sustainability mission statement or vision that defines overarching growth goals and commitments and communicates them throughout value chains and stakeholder communities.
- Develop a sustainability strategy emphasizing specific programs and defining opportunities in the value chain.
- Set time-based targets and goals and identify key performance indicators (KPIs) to track progress.
- Assign accountability for sustainability targets. Select the right team, business unit, or manager for each sustainability initiative tied to specific objectives and targets.
- Pursue sustainability initiatives: design and complete specific projects and initiatives to achieve accountable objectives and targets.
- Train staff on sustainability topics. Employees must understand how sustainability affects their industry, role, and organization's goals and operations.
- Decide on measurements and metrics essential to setting a baseline and tracking progress.
- Measure the sustainability impact of new projects. To assess new projects comprehensively, sustainability factors should be added to conventional financial risk and opportunity criteria.
- Educate stakeholders about the program and publish a sustainability report. Set stakeholder transparency goals and tailor information to each entity's needs in support of organizational change.
- Continuously monitor program activities. Sustainability programs are processes requiring constant refinement to ensure they reflect ever-evolving priorities.

Source: Rade, A., Ed., 27 March 2023, Workiva/Sustain.Life
www.sustain.life/blog/12-steps-corporate-sustainability-program
Retrieved: 11 March 2024

4.10.2 The Importance of Sustainability Metrics to Sustainability Management

The article noted that organizations operate in increasingly complex physical, demographic, political, social, cultural, technological, and economic conditions, often outside their ability to control or influence. However, the author advises that where the effects of these factors can be measured and managed, they should be.

Further, the article explains that the sustainability metrics discipline has evolved over the past decade from focusing on environmental concerns to broader social and organizational governance. Those social and governance concerns present measurement and management challenges far more complex than environmental sustainability, including the following.

- Broad social and governance concerns must be systematically measured and included with environmental metrics.
- ESG sustainability measures must be defined, analyzed, and prioritized to determine their contribution to financial success. They must be included with an organization's other KPIs, which will necessarily vary by organization.
- Designating sustainability metrics as KPIs within overarching management systems helps integrate sustainability into routine organizational administration.
- Developing systematic actionable sustainability metrics and analyses is necessary to meet current and future reporting requirements.

Source: Cohen, S., 11 December 2023, Columbia University Climate School

news.climate.columbia.edu/2023/12/11/the-importance-of-sustainability-metrics-to-sustainability-management/
Retrieved: 14 December 2023

4.10.3 The Current State of Advanced Information Technology Applications for Sustainability

BDA and AI promise broad, disruptive advances throughout societies, economies, technologies, and institutions. This, of course, includes advances in sustainability. Although practical applications are relatively new, these technologies are already entering maturation phases focused on business value rather than speculative future benefits. This bodes well for integrating BDA and AI into organizational sustainability as it, too, has matured to become more focused on business value.

The big question is whether these advanced information technologies are fulfilling their sustainability potential. Currently, the answer is no, not really, despite the efforts of early-to-market vendors. However, given marketplace needs and incentives plus the rate of technology development, they will soon.

Among its several points, a 2024 McKinsey article stressed that AI adopters:

- Use technologies that best serve the business in solving problems
- Understand and control cost factors
- Generate value safely and securely, and
- Provide the correct data, not necessarily the perfect data.

A *Harvard Business Review* article in 2023 prescribed five steps in AI application development and introduction for sustainability:

- Define sustainability objectives
- Centralize real-time and comprehensive data
- Customize current algorithms to unique sustainability challenges

- Pilot test in real-world conditions, and
- Full-scale implementation.

Several sources list these predicted and current sustainability BDA and AI applications:

- Sustainability program planning, performance tracking, and assessment, as well as ESG reporting
- Data capture, analysis, and reporting for regulatory requirements tracking and compliance
- Operations performance sensing, analysis, and reporting
- Closed-loop value chain management
- Waste management
- Water management
- Energy management
- Greenhouse gas management
- Materials development
- Product and service design, and
- Life cycle assessment.

Sources:

- www.mckinsey.com/capabilities/mckinsey-digital/our-insights/moving-past-gen-ais-honeymoon-phase-seven-hard-truths-for-cios-to-get-from-pilot-to-scale
- www.forbes.com/sites/forbestechcouncil/2023/11/22/14-ways-ai-can-help-business-and-industry-boost-sustainability/?sh=32048b86ed04
- hbr.org/2023/10/the-opportunities-at-the-intersection-of-ai-sustainability-and-project-management
- www.pwc.co.uk/sustainability-climate-change/assets/pdf/how-ai-can-enable-a-sustainable-future.pdf
- research.aimultiple.com/sustainability-ai/
- www.sciencedirect.com/science/article/abs/pii/S0040162522003262

Retrieved: 8 May 2024

The Sustainability Program Go-Decision and Policy Development

This chapter addresses the question in Chapter 4:

> *How will senior leadership – i.e. board of directors members and C-level offi-cers in the private sector and elected and appointed officials in the public sector – initiate a sustainability program and set its governing policies?*

It prescribes the critical early planning steps to create a sustainability program and assure executive buy-in. Specifically, it covers:

- The business case for sustainability
- Organizational change considerations
- The role of management system (MS) standards
- The need for a formal go-decision, and
- The importance of a foundational sustainability policy.

5.1 The Business Case for Sustainability

A sustainability program's go-decision initiates a major transformational change. Enthusiasm and altruism will only get sustainability champions so far in convinc-ing all levels of an organization to develop new ways to produce and deliver its products and services. The critical activity in operationalizing a sustainability pro-gram is the translation of environmental and social intentions into conventional organizational management terms. The core ideas in doing this are the 3P con-cept and the overarching *prime directive of finance* concept, commonly referred to as the *no money, no mission* reality.

The fundamental 3P concept suggests that organizations balance financial, social, and environmental performance. However, there is a compelling reason why the prime directive of finance, a crucial necessity, must take precedence

Sustainability Programs: A Design Guide to Achieving Financial, Social, and Environmental Performance, First Edition. William Borges and John Grosskopf.

over the other two. Altruistic and practical motivations for sustainability notwithstanding, every organization must recognize the no money, no mission reality. As crucial as environmental and social initiatives are, positive financial performance is a prerequisite to their consideration in decision-making. Of course, this requires a sophisticated, environmentally, and socially responsible approach to calculating *returns on investment* (ROI). With a more inclusive ROI approach, sustainability initiatives can be appropriately subjected to financial tests. It remains, though, that although environmental and social reasons may tip the scale for their adoption, ultimately, these initiatives must provide a suitable financial ROI. Otherwise, they will reduce the organization's financial performance and threaten long-term viability.

Of course, the no money, no mission concept is different at for-profit, non-profit, and public agency organizations. For example, because of the body of law regulating for-profit corporations, the standard view of the concept is this:

> *The only reason for-profit organizations exist is to increase the wealth of equity holders (e.g. proprietors and shareholders).*

It is capitalism at its best – or worst – depending on one's perspective. Unless a for-profit organization is financially healthy enough to produce profits and build up its balance sheet, it will not survive. In the business sense of the word, it will not be *sustainable*.

Nonprofits and public agencies also have their version of the prime directive:

> *Nonprofit and public-sector organizations must secure adequate financial resources to continue and possibly expand their missions.*

Although for-profit, nonprofit, and public agency organizations cannot be managed similarly, each must find its sweet spot in optimizing the 3P dimensions. Further, this sweet spot will constantly change as an organization's interests and conditions in its operating environment change. Therefore, this balance must be continually monitored, assessed, and adjusted in response.

Despite the dour-sounding no money, no mission reality, sustainability programs provide strong incentives for improving financial performance through greenwash-free environmental and social initiatives. Their key selling points are how they can help an organization systematically consider unrecognized opportunities for significant financial, environmental, and social ROIs. These include:

- Reducing and, in some cases, even eliminating environmental and social risks
- Reducing operating costs and administrative expenses, especially those associated with energy consumption and gaseous, aqueous, solid, and hazardous wastes

- Increasing revenues with innovative environmental- and social-attribute products and services, and
- Gaining tangible and intangible competitive advantages through transparency.

5.2 Organizational Change Considerations

The sustainability program approach described in this book is far from forming a few standing committees and *ad hoc* project teams to do occasional environmental or social-good projects. Instead, it is a systematic – and systemic – approach that not only achieves important environmental quality and social benefit goals but also financially benefits the organization. *Sustainability management systems* (SMSs) are the formal organizational structures and processes by which programs are implemented, operated, assessed, and improved.

Before making the all-important go-decision to create a sustainability program, the board of directors, the C-level officers, and other leaders must thoroughly understand that achieving sustainability requires *breakthrough performance* in risk management, cost and expense reduction, new revenues through innovation, and gains in competitive advantages. Breakthrough performance is a challenging proposition. It means that the *status quo* is no longer acceptable. Instead, only significant, measurable progress toward the best possible performance is. As a result, everyone must stop doing things in their professional and emotional comfort zones. Then, they must start doing unfamiliar things outside their current skill sets, interests, and emotional comfort zones. As difficult as it is, personal and organizational growth are the requisite change dynamics leading to breakthrough performance.

Change is scary and hard to do for many, if not most. People work long and hard to get through each workday with the least aggravation and the greatest personal reward. Once they reach their comfort zone, it can be difficult to dislodge them. It is no wonder that so many people react skeptically to organizational change and resist it. In extreme cases, some even sabotage change efforts. Therefore, it is critical that senior leadership formally prepare everyone for the upcoming changes well before they begin identifying, prioritizing, and responding to the organization's most pressing sustainability needs.

Two basic rules of life are:
1) Change is inevitable.
2) Everybody resists change.

W. Edwards Deming

CM efforts must explain, teach, encourage, and reward all employees to play their respective new roles in the SMS. Of course, expertise and related budgetary issues are always factors in providing adequate ongoing competency-focused instruction for new programs. This is especially true in organizations where under-valued *training-and-development* (T&D) functions are typically under-resourced and frequently targeted for cost-cutting. However, for complex initiatives like creating an SMS, such short-sighted thinking risks program underperformance or even failure at the outset.

In addition to teaching new processes and task responsibilities, CM efforts must consider individual and group motivations that promote willing and enthusiastic support for program success. A key concept is the tried-and-true *WIIFM* change motivation idea, i.e. *what's in it for me?* Financial incentives, formal and informal recognitions, prestige assignments, and promotions are critically important. Additionally, more subtle motivations are essential too. For some, it will be important to help the organization participate in fair trade programs, social justice efforts, charitable giving, and community volunteering. Whereas for others, participation in resolving environmental issues may be more important. In preparing employees for change, first, learn their concerns and interests. Then, tailor the CM activities to address those personal concerns and interests that align with the organization's.

CM efforts must stress that the organization will become a more effective enterprise by systematically defining and acting on the opportunities inherent in environmental and social challenges. The central theme of these efforts is the idea that the sustainability program is a critical part of the organization's mission, vision, and values where:

> *Well-formulated sustainability initiatives are not additional burdens on an organization. Instead, they are responses to heretofore unrecognized opportunities to manage risks, reduce costs and expenses, increase revenues through innovation, and gain other tangible and intangible competitive advantages.*

Reinforcing this theme, leaders must proactively and continually communicate the program's essential features, specifically its policies, plans, processes, activities, and expected and actual performance results. As previously mentioned, doing otherwise risks employee skepticism, resistance, and even program sabotage, leading to underperformance and possible failure. Chapter 3 and Appendix B, plus subsections in Chapters 1 and 4, provide additional CM considerations.

5.3 The Utility of ISO and Other Management System Standards

Since the 1980s, ISO MS principles and practices have been essential in helping organizations embed systemic improvements into their approaches to managing organization functions. Of the dozen-plus ISO management system standards, guidance, and guidelines in current use, the most relevant to sustainability are:

- *ISO 9001 Quality Management*
- *ISO 14001 Environmental Management*
- *ISO 450001 Occupational Health and Safety Management*
- *ISO 50001 Energy Management*

- *ISO 26000 Guidance for Corporate Social Responsibility*, and
- *ISO/UNDP PAS 53002 Guidelines for Contributing to the United Nations Sustainable Development Goals*.

As a social-performance-oriented standard, the *Social Accountability International* (SAI) *SA8000 Social Certification Standard* complements ISO's standards, guidance, and guidelines to round out topical certifications for a comprehensive SMS.

The SMS design process described in this book uses many of the same fundamental ISO MS concepts. The standards' vital features are their prescriptions for system documentation, administration, CI enhancements, performance tracking, improvement, and reporting. Using these prescriptions early in the SMS's development can save considerable effort later if the organization decides to pursue any certifications to the standards above.

Although they are discrete standards, ISO 9001, 14001, 45001, and 50001 are routinely integrated into a single MS along with the ISO 26000 guidance and SAI SA8000 as a best practice at the more proactive organizations. The ISO/UNDP PAS 53002 guidelines introduced late in 2024 add considerable detail to the topical focus of an SMS. Standards integration eliminates the administrative redundancies and other inefficiencies associated with individually adopting these overlapping standards.

Although industry MS standards have proven beneficial, are certifications confirming conformance with them necessary for the successful operation of an SMS? Consider this: performance improvements gained from ISO and ISO-like standards are independent of certification status. Thousands of organizations have demonstrated that considerable benefits accrue whether or not a program is ever certified.

Any tangible benefits of certifying an SMS to ISO standards must be weighed against the additional costs, administrative burdens, certification audit disruptions, and other distractions that divert attention and resources from creating and running a successful SMS. Therefore, unless a compelling reason exists, an organization using ISO standards in its SMS should consider the business benefits before seeking any certifications. Resources diverted to a certification process can be better employed to improve administration and operations-level sustainability performance. However, if an organization decides certification is beneficial, it would be less of a burden after implementing a well-documented and functioning SMS.

5.4 The "Go-Decision" to Create the Sustainability Program

In creating any new program that affects the overall performance of an organization, senior leadership must make a well-reasoned and documented decision to proceed. In other words, it must make a formal, unambiguous, compelling, and assertive *go-decision*. Regarding major initiatives, such as sustainability programs, the board of directors should take the lead through a clearly stated resolution or similar written mandate. Then, through the chief executive officer (CEO), the SMS champion, and the SMS oversight group – whose responsibilities are defined in Chapter 6 – this formal decision-to-proceed must be communicated along with the business case for sustainability to the organization's internal and external stakeholders.

Why is the go-decision so important? It communicates top management's mandate for the SMS initiative to everyone in the organization, helping secure its buy-in. Also, formal decisions to proceed by the highest levels of leadership are necessary because new major programs must compete for valuable resources needed by existing organizational units, functions, and departments. Senior leaders must ensure that competing units, functions, and departments continue to receive adequate resources to complete their work while achieving new sustainability objectives. To achieve this, the formal go-decision announcement must state that additional high-level sustainability considerations will be included in resource allocation criteria and, ultimately, how the organization will be run.

5.5 The Overarching Sustainability Policy

A formal high-level policy statement is necessary for any successful sustainability program. Of course, the opportune time to set this policy is when the board of directors issues its go-decision resolution.

The sustainability policy must unambiguously define guiding principles, programmatic objectives, and general requirements for the whole organization. The policy must also state the priority for these mandates relative to the organization's other declared intentions and requirements. These considerations set the depth and breadth of the organization's sustainability concerns, intentions, and limits.

In developing a sustainability policy, there are advantages to examining other organizations' sustainability programs – aka *benchmarking* – not least of which are the years and sometimes decades of experience gained from successful program management. Further, because many of these programs have excellent corporate transparency elements, their policies are readily available via the Internet. So, when drafting the sustainability program policy statement for a board resolution,

> *Observing many companies in action, I am unable to point to a single instance in which stunning results were gotten without the active and personal leadership of the upper managers.*
>
> **Joseph M. Juran**

it is not necessary to reinvent the wheel. Search the phrase *corporate sustainability policy*. Find those benchmarks that align with the organization's sustainability intentions and its mission, vision, and values.

In addition to benchmarking, an initial materiality assessment can narrow policy focus to topics most relevant within an organization's industry and its equity holders and other stakeholders. Materiality assessment is a standard needs assessment method favored by publicly traded corporations and the financial community. In business management, materiality refers to the issues most important to an organization. However, with sustainability programs, it is essential to distinguish between the terms *financial materiality* and *sustainability materiality*. In financial reporting, information is material if its omission or misstatement could influence economic decisions based on financial statements. In contrast, sustainability materiality refers to issues that may have significant repercussions; however, no formal monetary threshold has yet been defined to determine their financial materiality. As with any needs assessment, some companies retain external consultants while others manage the process themselves.

After defining and refining the organization's essential sustainability policy ideas, the go-decision proclamation and policy statements can be drafted for the board of directors' review, amendment, and adoption, followed by C-level officer implementation.

Note that some poorly conceived policy statements found online may have co-mingled overly specific or misguided strategic goals and tactical objectives and targets. The corporate sustainability policy is an essential overarching *grand strategy* statement. A grand strategy broadly states the approach that will be used to achieve long-term goals. Although it is potentially lengthy, it should be concise. Grand strategy documents are usually less than a page. So, resist the urge to include prescriptive details in the policy statement. To achieve brevity and program flexibility, it is best to leave the definition of specific goals, objectives, targets, and other administrative and operational requirements for later stages in program development.

Irrespective of benchmarked, assessed, and original content, the concise policy should be written in the context of the *sustainability efficiency* concept, i.e.:

> *Achieve the organization's people, planet, and profit goals with the least cost, effort, and risk throughout the entire closed-loop life cycles of its products and services.*

Although the efficiency concept requires that sustainability aspirations be achieved with the least cost, effort, and risk, it does not preclude them from being set at the highest breakthrough-performance levels. Of course, they can be set at low levels and every level in between. Even if they are initially set low, experience over time will help leaders decide how and when to raise aspirations. The expectation is that CI processes will eventually improve performance over time. This is a critical consideration for leaders at resource-strapped organizations.

An SMS's topical scope and organizational scale are essential considerations in the go-decision and policy development. *Chapter 7, The Sustainability Management System's Initial Needs Assessment,* discusses these factors further. Whatever the depth and breadth of a sustainability program's scope, it must be stated in the go-decision and the associated sustainability policy. This policy must also provide the program's critically important limiting boundaries. There should be no ambiguity around how much to do – or *not do* – to achieve the organization's sustainability goals. Clarity of the sustainability program's intentions must extend from the organization leaders to the employees and stakeholder communities.

Lastly, an organization must continually assess and refine the scope of its program in response to constantly changing marketplace conditions. Assessments and refinements are integral to the quarterly and annual SMS reviews described in Chapters 10 and 11.

All of this leads to an obvious question:

Who writes the policy?

Drafting the sustainability policy for board of directors deliberation, amendment, and acceptance is a terrific opportunity for the C-level officers as agents of the board to involve individuals at all levels of the organization in a significant *shared governance* activity. The advantages of shared governance at this program development phase are early buy-in, continued support, and promotion of cross-functional collaboration at all levels. The importance of buy-in, support, and collaboration from the start cannot be overstated when effectively managing a program's inevitable changes to an organization's administration and operations.

5.6 Relationships Between Sustainability Policy and Mission, Vision, and Values Statements

At their simplest, an organization's *mission, vision,* and *values statements* proclaim:

- Why the organization currently exists
- The organization's intentions for future development, and
- The moral, ethical, and legal principles it observes.

A new sustainability policy can amplify, influence, or even conflict with an organization's mission, vision, and values. In such cases, sustainability policy development must consider these three statements. Notably, after the policy is adopted, it may be necessary to amend one or more statements for consistency immediately or during the next strategic planning cycle.

5.7 Chapter Takeaways

- The essential tactic in operationalizing a sustainability program is the translation of environmental and social intentions into conventional organizational management terms.
- When building a sustainability program, organizations must recognize the prime directive of finance's no money, no mission reality. Without adequate financial performance, there will be no organization that can achieve environmental and social aspirations.
- Sustainability programs help an organization systematically consider formerly unrecognized opportunities for significant financial – as well as environmental and social – ROIs, including:
  - Reducing and, in some cases, eliminating risks associated with global, regulatory, and industry standards for financial performance, contractual obligations, product and service quality, social responsibility, and environmental stewardship
  - Reducing operating costs and administrative expenses, especially those associated with energy consumption and gaseous, aqueous, solid, and hazardous wastes
  - Increasing revenues with innovative environmental- and social-attribute products and services, and
  - Gaining tangible and intangible competitive advantages through transparency.
- Each organization must find and maintain its sweet spot in balancing the 3P performance dimensions. This balance constantly changes as interests and operating environment conditions change. Therefore, it must be continually monitored, assessed, and adjusted in response to those changes.
- Achieving sustainability improves organizational performance – including breakthroughs – in risk management, cost and expense reduction, new revenues through innovation, and gains in competitive advantages via transparency.
- Effective sustainability programs require structured, well-resourced CM efforts.
- ISO and other MS standards can be useful aids to the successful design and development of sustainability programs and their management systems.
- The go-decision to create a sustainability program must be made formally by the board of directors for implementation by C-level officers.

- The board of directors is responsible for adopting a high-level sustainability policy implemented by the C-level officers that addresses financial, environmental, social and other organizational objectives.
- The sustainability policy should be brief and include the sustainability efficiency concept:

> *Achieve the organization's people, planet, and profit goals with the least cost, effort, and risk throughout the entire closed-loop life cycles of its products and services.*

- The sustainability policy should be drafted by a group of C-level officers with the support of select individuals from all levels of the organization. This shared governance activity encourages early buy-in, continued support, and cross-functional collaboration at all levels.
- Sustainability policy development must consider an organization's mission, vision, and values statements. After policy adoption, it may be necessary to amend these statements for consistency immediately or during the next strategic planning cycle.

5.8   Further Reading

Supporting the book's ideas are three discussions on:

- The business case for sustainability
- The definition and features of business policy, and
- The integration of various MS standards.

5.8.1   Making the Business Case for Sustainability

This blog post describes eight benefits of a sustainable business strategy, consistent with the book's content. Further, the author argues that sustainability initiatives should not be considered financial trade-offs but wise financial strategies.

- *Sustainability drives internal innovation*: Switching to sustainable business practices provides opportunities for new, innovative ideas to achieve business growth, energy savings, and ethical material sourcing benefits. More broadly, sustainability can stimulate ideas for new business opportunities.
- *Sustainability improves environmental and supply risk management*: Sustainable practices can improve risk management in several areas, citing upstream energy supply chain security and cost control as examples.

- *Sustainability attracts and retains employees*: Sustainability commitments display organizational values, attracting job seekers who share those values. Hiring and retaining the right team saves organizations time and money by preventing the need to rehire constantly. Citing a large-scale survey, the author notes that:
  - A substantial majority of employees are influenced by sustainability commitments to stay with a company long-term
  - Most millennials are willing to take a lower salary for the opportunity to work at an environmentally responsible company, and
  - Nearly half of respondents took one job over another because of the organization's sustainability practices.
- *Sustainability expands audience reach and builds brand loyalty*: Adopting sustainable practices and marketing appropriately can help an organization reach new, sustainably minded market segments while building brand loyalty. The post notes that sustainable businesses see more significant financial gains than their unsustainable competition.
- *Sustainability reduces production costs*: The blog notes that production efficiency and lower costs are incentives for sustainability since fewer resources are used and production inputs replace more impactful ones.
- *Sustainability garners positive publicity*: Positive publicity around the sustainability transition can benefit employee satisfaction, job applicant attraction, and customer loyalty.
- *Sustainability helps a company stand out in a competitive market*: Sustainability can be a positive differentiator from competitors. The author notes that consumers' focus on brand sustainability practices could be the sole reason they choose a product over a competitor's.
- *Sustainability sets the industry trend*: Widespread corporate adoption of sustainable practices has the potential to impact the world's most significant problems. It also helps companies stand out from competitors, influencing their behaviors and establishing themselves as trend-setting leaders that prompt other companies to follow.

Source: Cote, C., 13 April 2021, Harvard Business School Online Business Insights Blog

online.hbs.edu/blog/post/business-case-for-sustainability
Retrieved: 12 March 2024

5.8.2 Business Policy – Definition and Features

Business education portal *ManagementStudyGuide.com* provides a detailed outline and explanations of business policy development considerations, including:

- Definition of policy
- Features of policy

- Criteria for effective policies, and
- Distinctions between policies and strategies.

Source: Managementstudyguide.Com
managementstudyguide.com/business-policy.htm
Retrieved 3 March 2024

5.8.3 ISO Handbook: The Integrated Use of Management System Standards

Recognizing the efficiency advantages of integrating several MS standards into an organization's overall MS, ISO produced *The Integrated Use of Management System Standards* (IUMSSs) handbook, which was last revised in 2018. Divided into three chapters, the handbook describes:

- The fundamentals of MSs and how they link an organization's strategies, plans, and operations
- The structures and contents of various MS standards and their applications, and
- How organizations can integrate the requirements of multiple standards into their overarching MS.

Source: The International Organization for Standardization, 2nd Ed, November 2018
https://www.iso.org/files/live/sites/isoorg/files/store/en/PUB100435_preview .pdf
Retrieved: 14 May 2024

6

A Model Sustainability Management System Organization Structure

This chapter details the SMS model organization structure and administrative processes, and the ways they depend on and integrate with existing management structures and functions. The organization structure illustrated in Figure 6.1 enables capability creation and performance improvement outcomes with the least cost, effort, and risk. Although this chart assumes a moderately large enterprise, it can be scaled up or down to fit any organization.

The following questions from Chapter 4 guide the structuring of an SMS and address various ways to assure executive buy-in.

- *How will the SMS's organizational structures, including functions, lines of authority, and stakeholder roles, be designed?*
- *Who among the board of directors members, C-level officers, other executives, or senior public sector officials will be the SMS champion?*
- *How will executive, managerial, and other participant accountabilities for SMS success be set and incentivized?*

Figure 6.1 helps answer these questions by illustrating the SMS's roles, responsibilities, and lines of authority between board of directors members, C-level officers, the SMS oversight group, the SMS champion, supporting functions and departments, SSTs, and *ad hoc* project teams.

6.1 The SMS's Organization Structure

The SMS creation efforts described so far have involved leaders and some staff members within the existing organizational structure. However, to effectively manage an SMS over the long term, it is necessary to define and implement distinct:

- Management system functions
- Roles, responsibilities, and lines of authority within those functions

Sustainability Programs: A Design Guide to Achieving Financial, Social, and Environmental Performance, First Edition. William Borges and John Grosskopf.
© 2025 John Wiley & Sons, Inc. Published 2025 by John Wiley & Sons, Inc.

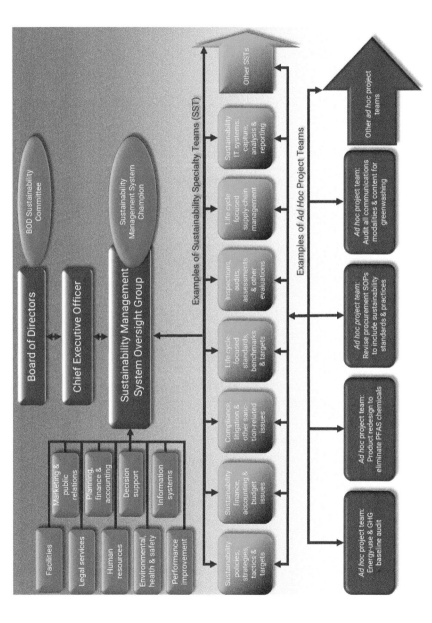

Figure 6.1 A Model SMS Organization Chart.
Source: W. Borges.

- Work processes, and
- Performance accountability mechanisms.

It is important to emphasize that these SMS elements do not require an organization – especially small- and mid-sized ones – to create a stand-alone sustainability department. However, leaders may create one at some point in the SMS's development. This is especially true of larger, complex, and more sophisticated organizations. Remember, there are no threshold criteria triggering the creation of a stand-alone department. That case-by-case decision will also consider the prospective department's appropriate size, structure, and responsibilities.

From essential management *planning, organizing, controlling,* and *leading* perspectives, it is essential that leaders and staff know on an ongoing basis:

Who is accountable for what, and by when?

The *who* part of this question is the critical consideration in this chapter. Ultimately, an SMS design must be tailored to unique sustainability needs, especially the organizational structure that supports it. The model organization chart in Figure 6.1 shows various SMS roles and lines of authority. Further, typical responsibilities for each role are described below. It is suggested that the SMS designers figuratively *beat the stuffing out* of the model and put it back together in whatever way works best for their organization. In doing so, SMS designers must consider the adequacy of existing personnel resources versus the need for outside assistance or additional hires.

6.2 Leadership Roles in the SMS

Here are the typical SMS leadership roles based on current examples of successful sustainability programs in the United States. The roles and responsibilities in this SMS organization model are common in conventional private sector corporations. However, titles and responsibilities may need to be adjusted in privately held small firms, nonprofits, and public agencies. When adjusting the model, remember that a key to SMS success is the definition of clear lines of authority and concise communication processes between each of these roles:

- Board of directors and its sustainability committee
- Chief executive officer (CEO)
- SMS oversight group, and
- SMS champion.

Because of its importance, the SMS champion role is described first in the following text despite its position on the Figure 6.1 organization chart.

6.2.1 SMS Champion

Far too many organizations form an *ad hoc* sustainability committee or appoint a low-level coordinator or manager to do environmental, social-benefit, or cost-saving projects. These usually focus on the short term and are typically ineffective in achieving substantial sustainability goals over the long term.

In contrast, the most effective sustainability programs are led by an SMS champion. SMS champions are critical because they are the administrative control points for enterprise-wide sustainability programs. Subordinate champions may be needed in large organizations with multiple strategic business units. Also, due to their values, interests, insights, and responsibilities, there may be several informal champions, especially in larger organizations. However, only one person should be accountable for managing the overall program.

In their leadership role, an SMS champion is the principal program designer, technical expert, change agent, controller, facilitator, mediator, educator, mentor, coach, cheerleader, and, when necessary, disciplinarian up and down the organization chart. They are also responsible for recruiting personnel to participate in SMS activities. These responsibilities must be considered when deciding who the SMS champion reports to. In addition to battle-tested sustainability expertise, this kind of responsibility requires significant, recognized, and respected *position, referent, information, coercion, reward*, and other leadership authorities in the organization. Typical titles reflecting such authority include:

- Chief sustainability officer (CSO)
- Vice president for sustainability, or at the very least
- Corporate sustainability director.

Position titles to avoid for the SMS champion are *specialist, coordinator,* and *manager*. Why? Because SMS champions have transformational responsibilities, whereas specialists, coordinators and managers with their transactional ones rarely have the leadership clout, authorities, and change-agent skills required to lead major new initiatives. Further, the reality is that few board members, C-level officers, vice presidents, other executives, or their pubic sector counterparts will effectively respond to the earnest efforts of hard-working, lower-level personnel.

That is why companies with effective sustainability programs have CSOs, sustainability vice presidents, and corporate sustainability directors. They recognize that sustainability is a significant organizational management function that demands strong leadership, experiential, and intellectual horsepower to measurably achieve an organization's *people, planet,* and *profit (3P)* goals with the least cost, effort, and risk.

If the SMS champion's title is less than CSO, their accountabilities should be drafted by the SMS oversight group for the CEO's review, amendment, and assignment. Whereas if the SMS champion holds the title of CSO, their accountabilities should be defined and assigned by the board of directors.

These titles warrant significant compensation in recognition of the job security and other career risks inherent in creating and growing a major transformational program. Effective sustainability programs mean that other leaders and staff will likely have to be moved out of their professional and emotional comfort zones. The SMS champion's compensation must be worth the position's effort, risks, and associated stresses.

6.2.2 Board of Directors and Its Sustainability Committee

Relying on the recommendations of its sustainability committee, the entire board of directors has ultimate *environmental, social, and governance* (ESG) authority over a sustainability program and its SMS. Depending on the size of an organization and the number of directors, a sustainability committee may include the entire board or a subset of its membership. The board of director's responsibilities include:

- Initially evaluating whether the organization should implement a sustainability program and its associated management system, i.e. the *go/no-go decision*
- Assuming a go-decision, adopting and periodically revising the organization's high-level sustainability policies – including the guiding principles, programmatic intentions, and other general requirements – to achieve 3P aspirations with the least cost, effort, and risk throughout the entire closed-loop life cycles of its products and services
- Monitoring the organization's overall sustainability progress and prescribing policy-level enhancements and corrective actions as needed
- Defining the incentive compensation program and assigning the associated sustainability accountabilities for all CxOs, and
- Providing monitors to the SMS oversight group.

> **A Note Regarding SMS Incentive Programs**
>
> The board of directors will typically design incentive programs for the CxO leadership based on established practices. Then, the CxO leadership team will draft incentive programs for board review, amendment, and approval for lower-level executives, managers, and other SMS participants.

6.2.3 Chief Executive Officer

The CEO is the board of directors' primary agent for implementing policies at the organization's administration and operations levels. As such, the CEO is responsible for setting priorities and marshaling resources to create new capabilities and resolve performance issues promptly and cost-effectively. With these responsibilities, it is easy to see how vital the CEO's involvement and commitment are to the success or failure of a sustainability program.

However, with all the competing demands for the CEO's time and attention, the detailed SMS work must be delegated to others who will provide them with the right information to make the right decisions at the right times. In this case, the delegated parties are members of the SMS oversight group, which is administered and often led by the SMS champion.

6.2.4 Sustainability Management System Oversight Group

The SMS oversight group manages the SMS's programmatic operations. Its administrative focus is on assuring and improving the overall ability of the SMS to achieve the organization's sustainability goals with the least cost, effort, and risk within its resource limits and opportunities. Its operational focus is assigning, tracking, and evaluating capability creation and performance improvement projects to meet the organization's high-level, time-constrained strategic goals and specific tactical objectives and targets.

As such, it is not a technical group. Instead, in its program management role, it recommends capability creation and performance improvement initiatives to the CEO that should be undertaken in the organization's administration and operations units, functions, and departments. Once the CEO reviews, revises as necessary and accepts those recommendations, the SMS oversight group monitors overall program efficacy, tracks and supports individual project progress, and evaluates successes and failures.

The SMS oversight group's composition may vary according to the organization's size and complexity. The SMS oversight group typically consists of five to seven high-ranking leaders in larger organizations. The titular head of the group should be the CEO. However, the CEO may need to delegate group leadership responsibilities to one of their direct CxO reports or the SMS champion. As noted earlier, the board of directors should appoint one or more monitors to the group to ensure program success. Members should be drawn from the ranks of senior leadership, such as the:

- Chief Operations Officer
- Chief Financial Officer
- Chief Administration Officer
- Chief Technology Officer
- Chief Information Officer
- Chief Strategy Officer
- Chief Marketing Officer

- Legal Affairs Director
- Decision-Support Director
- Performance Improvement Director
- Facilities Director
- Risk Management Director
- Environmental, Health and Safety Director, and
- Human Resources Director.

Of course, a permanent group member is its administrator, the SMS champion. For other membership terms, a best practice is a minimum of one year. Further, the start and end dates of the members' terms should be staggered to ensure continuity of experience and leadership within the group.

Reporting to the CEO, the SMS oversight group's primary responsibilities should include the following.

- Complete the SMS's initial needs assessment to define the organization's most pressing sustainability needs. See Chapter 7.
- Draft for the CEO's approval:
  - Periodic – at least annual – recommendations for new and revised strategic goals, as described in Chapter 8
  - Annual tactical objectives and targets, as described in Chapter 9
  - SMS administration processes, and
  - Any necessary administrative or operational modifications to the SMS based on its annual review, as described in Chapter 11.
- Draft quarterly priorities and associated achievement schedules for strategic goals, tactical objectives, and targets for CEO approval, as described in Chapter 9.
- Before the start of each fiscal quarter, define the event calendar for the next 13 weeks. See Chapter 10.
- Demonstrate in scheduling the accomplishment of sustainability goals, objectives, targets, and initiatives that adequate consideration has been given to the organization's administration and operations activities and other conventional capability creation and performance improvement initiatives. Then, assign specific unit, function, and department-level accountabilities to complete sustainability-themed capability creation and performance improvement projects. See Chapter 10.
- Recruit members for the SSTs and assign and track their project-support accountabilities. Refer to the description of SSTs later in this chapter.
- Assure that capability creation and performance improvement projects have adequate resources to succeed, especially regarding program-participant workloads, budgets, and other organizational resources. See Chapter 10.

- Periodically monitor project progress to successfully achieve the capability creation and performance improvement account-abilities on time and within budgets. See Chapter 10.

> *A bad system will beat a good person every time.*
>
> **W. Edwards Deming**

- Formally assess projects and celebrate their success, ideally at the end of each fiscal quarter. Or, in the case of unsuccessful projects, define the root causes of failure and take effective corrective actions *without punishing people for the sins of the work process.* The exception to this admonishment is, of course, where punitive personnel actions are warranted. This last point about punishing people is critically important because project failures often result from faulty work processes and associated support systems. Thus, if the SMS is to be accepted throughout the organization as a beneficial initiative, scapegoating in its many forms must be avoided. See Chapter 10.
- Participate in communications activities with internal and key external stakeholders regarding SMS performance and the accomplishments of key contributors.

As required, advise and direct the SMS champion on any actions necessary to ensure the SMS's ongoing success.

- As described in Chapter 11, conduct an annual full-scope management review of the SMS to ensure its continuing suitability and effectiveness. The review should focus on the:
  - Current efficacy of the overall SMS with an emphasis on identifying specific inabilities to achieve objectives with the least cost, effort, and risk
  - Successes, failures, and lessons learned from the current portfolio of capability creation and performance improvement projects
  - Status of actions taken in response to previous management reviews
  - Adequacy of current SMS resources and expected needs for new ones
  - The necessity for new needs assessments – e.g. *strength-weakness-opportunity-threat* (SWOT) analyses, materiality assessments, life cycle assessments, formal audits, and stakeholder surveys – especially in consideration of the findings from other evaluations completed since the last SMS review
  - Changes in the status of the organization, including:
    - New or increasing internal and external risks and opportunities
    - Internal administrative and operational changes, and
    - Other change-related opportunities for improving the SMS.
  - The production of a report summarizing the SMS's most pressing needs, *opportunities for improvement* (OFI), and specific actions required to ensure the continued efficacy of the SMS to add value to the organization. It is important to note that this annual SMS evaluation report is an essential informaton source for any voluntary or mandated ESG reporting activities.

6.3 Support Roles for the Sustainability Management System

The SMS oversight group assigns and tracks accountabilities for various support functions, especially the ones described below. Although the following functions are usually organized as departments, many are not in some smaller organizations. That is why they are referred to here as functions.

6.3.1 Organizational Planning, Finance, and Accounting

The organizational planning, finance, and accounting functions support the SMS by incorporating sustainability as a critical element in the organization's overall strategic, tactical, and budgetary planning and control processes at the unit, function, and department levels.

6.3.2 Decision Support

In many industries, organizations have discrete decision-support departments. However, in organizations lacking them, various information capture, analysis, and reporting departments collectively provide this function. The decision-support function, however it is constituted, supports the SMS by incorporating sustainability performance factors into its respective activities. These activities should minimally include sustainability-focused:

- Historical and current financial and managerial accounting analyses
- Stakeholder satisfaction tracking and reporting
- Trend forecasting, and
- Other special analyses.

Sustainability *key performance indicators* (KPI) based on strategic goals, tactical objectives and targets, and unit, function, and department accountabilities must be included in periodic dashboard, balanced scorecard, and other performance reports. The decision support function also assists with internal and external SMS communications activities.

6.3.3 Information Systems

Working with the SMS oversight group and the other support functions, the organization's information systems function proactively and continuously participates in the SMS. It automates as many information capture, analysis, and reporting activities as possible within its resource limits and opportunities.

6.3.4 Legal Services

In sustainability programs, the legal services function adds new, complex, and evolving risk management concerns to its conventional law, policy, administration, operations, and investor relations responsibilities. Some of those concerns are:

- ESG reporting
- Sustainability laws and regulations
- Climate change and environmental degradation regulatory compliance
- Legal and regulatory compliance with internal and external social responsibility requirements
- Due diligence support to identify ESG issues in transactions, value chains, and high-risk projects
- Strategy and tactics advice on external environmental and social responsibility stakeholder relations, and
- To prevent greenwashing liabilities, review and advise on external communications, including financial and ESG reporting, press releases, public relations (PR) activities, and marketing media.

6.3.5 Human Resources

The human resources (HR) function is essential to the design, implementation, and management of an SMS. Its responsibilities typically include:

- Advising on compliance with sustainability's internal and external social responsibility standards during organization policy development and periodic identification of OFIs
- Organizational development support
- Change management planning and implementation
- Training and development services
- Participation in developing and administering personnel incentive programs
- Recruiting and onboarding sustainability program personnel
- Identifying internal personnel for participation in the SMS, and
- Internal SMS communications.

6.3.6 Environmental, Health, and Safety

Within sustainability's broad 3P concepts and disciplines, the *environmental, health, and safety* (EHS) function provides foundational regulatory compliance services. These services are focused on the health and safety of employees and external stakeholders while protecting the environment from operations-related impacts. The function's extensive expertise with treaties, laws, and regulations, and its experience in defining and refining processes to monitor, detect, and correct regulatory variances and nonconformances are invaluable components of an SMS. Specifically, the expertise of EHS staff helps define operations-level OFIs and assists SSTs and *ad hoc* project teams in designing and completing capability creation and performance improvement projects.

6.3.7 Performance Improvement

Dedicated to eradicating value chain wastes of all types, performance improvement functions have decades of experience with CI methods, including Lean Manufacturing, Six Sigma, and Total Quality Management. The function's experts use these and other systematic approaches to identify operational issues, define root causes, plan effective corrective actions, and coach personnel through to successful problem resolution. The performance improvement function's methods and expertise are vital in supporting SSTs and *ad hoc* project teams in successfully designing and completing capability creation and performance improvement initiatives. In addition to its *ad hoc* project-support activities, the function assists with overall SMS improvements required by the annual review detailed in Chapter 11.

6.3.8 Facilities

Sustainable facilities management considers the life cycle of a facility from design and construction to operation and maintenance to end-of-life disposition. It is important to note that the U.S. Green Building Council's *Leadership in Energy and Environmental Design* (LEED) program concentrates on sustainable facilities management. Facility managers focus on efficiently using an organization's physical plant assets. Their primary concerns include:

- Facility design, construction, maintenance, and final disposition processes
- Energy management
- Waste management
- Water management, and
- Expense and cost containment.

6.3.9 Public Relations and Marketing

Public relations (PR) and marketing functions support SMS leaders and other departments with advisory services to effectively communicate the sustainability program's policies, goals, and performance outcomes. However, a cautionary note is warranted. Even though PR and marketing departments may seem well suited, they cannot be given the chief responsibility for the program's reporting and other communications activities. This is because the functions of these departments are to promote an organization's products, services, and overall image to gain tangible and intangible marketplace advantages. Due to this promotional bias, i.e. shining the best light on products, services, and the organization, these functions cannot be relied on to assess and report on a sustainability program objectively. It is essential to recognize these functions' paradigms when reporting on baseline sustainability needs and measured progress to meet them in ways acceptable to critical regulators, investors, and other stakeholders. Further, when they are unchecked, PR and marketing functions often create unnecessary greenwashing risks. Therefore, responsibility for sustainability program reporting, an essential ESG activity, must reside with the SMS champion, SMS oversight group, CEO, and ultimately, the board of directors.

6.4 The Capability Creation and Performance Improvement Roles in the SMS

The following bullet points describe the working relationships between the SMS oversight group administered by the SMS champion, the SSTs, and the *ad hoc* project teams. Together, they are responsible for implementing the SMS's cascading policies, strategies, tactics, and targets via capability creation and performance improvement initiatives. These shared governance relationships are essential to SMS success.

- The SSTs produce needs assessments and subsequent updates, as described in Chapter 7. Each SST assesses, defines, and prioritizes the organization's strengths, weaknesses, opportunities, and threats around one and sometimes more specific sustainability topics. The topics that might be assigned to an SST are shown in Figures 7.1–7.4. These needs assessments use information from the SMS support functions, various administration and operating units, focused research, current industry and competitive trends, and team members' knowledge.
- In turn, on behalf of the SMS oversight group, the SMS champion aggregates and shortlists prioritized sustainability needs identified by the SSTs, as described in Chapter 7. These are now the organization's most pressing sustainability needs. In this aggregation process, the sustainability needs are described as OFIs from here on out. The OFIs are used to define strategic goals and tactical objectives and targets, as described in Chapters 8 and 9.
- Regardless of current budgetary constraints, the SMS oversight group recommends to the CEO the OFIs that best conform to time requirements established

by strategic goals and their associated tactical objectives and targets in the form of capability creation and performance improvement accountabilities. Any unassigned OFIs are listed separately and saved for future consideration. See Chapter 7.

- When deciding which OFIs to exploit during the next improvement cycle, the CEO considers:
  - Sustainability policies, goals, objectives, and targets;
  - Resource constraints and opportunities; and
  - Other relevant factors.

- With the development of strategic goals and tactical objectives and targets completed, as described in Chapters 8 and 9, the SMS oversight group assigns capability creation and performance improvement accountabilities approved by the CEO to specific organization units, functions, or departments. See Chapter 10.

- Capability creation and performance improvement projects are designed and completed by *ad hoc* project teams created by the accountable unit, function, and department leaders. With their topical expertise, the SSTs support the *ad hoc* project teams with design assistance. See Chapter 10.

- During the conduct of these projects:
  - The SSTs have project-specific progress tracking and support responsibilities
  - The SMS champion monitors project progress and supports the SSTs as needed, and
  - The SMS oversight group has program administration and support responsibilities. See Chapter 10.

6.4.1 Sustainability Specialty Teams

Typically consisting of two or three people each, the SSTs focus on one – or possibly more – of the sustainability topics shown in Figures 7.1–7.4. Ideally, the SST members should have expertise related to the team's topic. However, instead of expertise, especially in organizations with limited resources, a strong interest and enthusiasm will suffice during the SMS's early development phases. With time, experience, and strong management support, newcomers can develop the necessary expertise to contribute to SSTs' success. The sustainability learning resources in *Appendix C* can be helpful in building competencies.

The SMS oversight group with the help of the HR and other functions should actively recruit SST members as part of the organization's leadership development efforts and set and track their accountabilities. As with the SMS oversight group, SST membership terms should be at least one year. Further, each member's term start and end dates should be staggered on a fiscal quarter or biannual basis to assure continuity of expertise within the SST.

SST responsibilities include the following.

- During the SMS creation, the SSTs will define topic-specific sustainability needs using assessment results from assigned SWOT analyses. From these initial needs assessments, each SST will produce and deliver to the SMS oversight group a prioritized list of the organization's most pressing sustainability needs for its particular topic(s). See Chapter 8.

- Ideally, during each fiscal quarter following the initial creation of the SMS, the SSTs will use inputs from the organization's support functions, any new audits or assessments, project experiences, industry trends, and other performance-sensing resources to provide updated needs assessments. From these quarterly needs assessments, each SST will update the SMS oversight group with its prioritized list of the organization's most pressing sustainability needs for its particular topic(s). See Chapter 10.

- As described earlier, before each subsequent fiscal quarter ends, the SMS champion, on behalf of the SMS oversight group, will update its list of OFIs using the aggregated most pressing needs information produced by the SSTs. Then, the SMS oversight group will prioritize and recommend new projects to the CEO that best conform to the schedule to achieve strategic goals and associated tactical objectives and targets during the next fiscal quarter. See Chapter 10.

- Upon CEO approval of the project list, the SMS oversight group will assign specific unit, function, and department-level accountabilities. *Ad hoc* project teams will be formed in the accountable units, functions, and departments. Then, the associated SSTs will work with them and the SMS champion to design appropriate projects to achieve the assigned accountabilities. See Chapter 10.

- In situations requiring extraordinary resources for project success, the SSTs, with the assistance of the SMS champion, will present the justification along with a budget request, when necessary, to the SMS oversight group for action. See Chapter 10.

- Ideally, on a weekly or biweekly basis, the SSTs will review the *ad hoc* project teams' progress. When adverse variances and nonconformances to a project plan occur, the SSTs will assist *ad hoc* project teams in determining root causes and then track immediate and effective corrective actions to return projects to plan conformance. The SSTs will consult with the SMS champion and the SMS oversight group to resolve issues when necessary. See Chapter 10.

- The SSTs will provide formal end-of-quarter and end-of-project reports on their project portfolios to the SMS champion for use by the SMS oversight group. At a minimum, these reports will present objective evidence of project performance and outcomes. In doing so, the reports will discuss performance issues, the results of corrective actions, and prescriptions for the next steps. See Chapter 10.

As described in Chapter 10, *ad hoc* project teams are formed to successfully achieve tactical objectives and target accountabilities with the support of an assigned SST by completing a capability creation or performance improvement project approved by the CEO and assigned by the SMS oversight group. *Ad hoc* project teams may be formed in a single or across multiple units, functions, or departments. Acting on behalf of the SMS champion, associated SSTs will provide technical and other oversight support to the *ad hoc* project teams in their efforts. However, individual *ad hoc* project teams are ultimately responsible for project success.

Because of the variability of project types, the numbers and sizes of *ad hoc* project teams will vary. Further, the terms of *ad hoc* project team membership will be set by the length of the project and an individual team member's task assignments. Lastly, as an *ad hoc* project team, there are no ongoing participation requirements once the project formally ends. *Ad hoc* teams are disbanded once the SMS oversight group declares project completion as a part of a formal closure process.

Ad hoc project team responsibilities include the following.

- With the advisory assistance of the responsible SST on behalf of the SMS champion, each *ad hoc* project team will design a plan to accomplish its assigned accountability. Project planning details are provided in Chapter 10 and Appendix E. These project plans will include the following:
  – Project Overview
  – Task Descriptions
  – Project Schedule, and
  – Project Budget.
- Each *ad hoc* project team will work to successfully execute its plan to achieve its accountabilities on time and within budget.
- The *ad hoc* project teams will submit written weekly, biweekly, or monthly progress reports, as required, to their respective SSTs. The reports will include:
  – A summary of completed tasks
  – A description of any variances or nonconformances to the project plan, their root causes, and corrective actions
  – An assessment of the efficacy of corrective actions to project plan variances or nonconformances, and
  – A preview of the project's next steps, including anticipated challenges and measures to avoid or adequately mitigate those challenges.
- The *ad hoc* project teams will submit to their associated SSTs a closure report at the end of their projects, which will, in turn, be forwarded to the SMS champion. The report will include:
  – A summary of the work effort and the contributions of team members to meet their accountabilities;

- Objective evidence of project success or a root cause analysis of project under-performance or failure
- A summary of lessons learned, and
- Prescriptions for the next steps, including
  - Recommendations for sustaining the project outcomes and
  - New actions to resolve issues discovered during the project.

6.5 Chapter Takeaways

This chapter prescribes an SMS organization structure model that enables capability creation and performance improvement outcomes with the least cost, effort, and risk. It details:

- Management system functions
- Specific roles and responsibilities within those functions
- Lines of authority
- Work processes, and
- Performance accountability mechanisms.

The primary personnel roles in that structure are:

- SMS Leadership Roles
  - Board of directors and its sustainability committee
  - CEO
  - SMS oversight group
  - SMS champion
- Support Roles
  - Organizational planning, finance, and accounting
  - Decision support
  - Information systems
  - Legal services
  - Human resources
  - Environmental, health, and safety
  - Performance improvement
  - Facilities
  - Public relations and marketing
- Operations Roles
  - SSTs
  - *Ad hoc* project teams

6.6   Further Reading

Supporting the book's SMS organization structure ideas are three discussions on:

- Sustainability program structural elements
- The evolving role of CSOs, and
- The organizational structures at Verizon, GE HealthCare, and Walmart.

6.6.1   How to Build Effective Sustainability Governance Structures

Consistent with the book's premise that organizational structure drives behavior and, ultimately, performance quality and culture, BSR's blog post states that successful sustainability integration and effective management require committed leadership, clear direction, strategic influence, and a robust governance structure. Further, sustainability governance helps a company implement strategy, manage reporting processes, strengthen relations with external stakeholders, and ensure overall accountability.

The blog post also describes how and where sustainability fits into the overall structure and reveals an organization's direction and priorities. As does this book, the post cautions that there are no cookie-cutter, one-size-fits-all structures. Every organization must tailor its approach to fit its business model, structure, resources, and level of desired sustainability integration.

The blog recommends that the following concepts be considered when building a governance structure:

- Commitment begins at the top
- Accountability must be established and communicated clearly
- Alignment between the sustainability program structure and the business is imperative, and
- Flexibility to adapt and build up on the sustainability program across business units and regions can advance the sustainability agenda.

With these considerations in mind, BSR prescribes the following best practices.

- *Formal Board Committee*: Although board oversight of sustainability matters is typically performed across several formal committees, it can be accomplished through a dedicated committee. Committees are essential for educating boards on sustainability issues and demonstrating high-level commitments to sustainability.
- *Head of Sustainability*: Most of BSR's corporate clients have dedicated sustainability leaders with titles like CSO who demonstrate commitment, drive sustainability strategy, and advance program effectiveness.

- *Cross-Functional Executive Sustainability Committee*: Analogous to the book's SMS oversight group, a cross-functional, high-level executive committee provides further oversight and strategic guidance within operating organizations by engaging leadership across business units, regions, and functions while mobilizing employees to implement strategies. The various organizational functions include risk management, supply chain, operations and facilities, marketing, public affairs and communications, human resources, environmental, health and safety, and investor relations.
- *Sustainability Teams*: A core team of sustainability experts, such as a sustainability department, can help coordinate daily activities and implement companywide initiatives. Although such departments are common in large organizations, avoiding their siloing is essential when engaging with and integrating into business units and functions.
- *Sustainability Supporting Functions*: Akin to the book's SSTs and support roles of existing organizational functions and departments, working groups or committees with a reporting relationship to the head of sustainability can assist in implementing strategic goals and tactical objectives and targets. They can do so by supporting and, in many cases, substituting for expert sustainability teams. Individuals in these support structures may come from real estate, facilities, communications, human resources, risk management, supply chain, and other groups. They may be accountable for priority sustainability topics, strategy implementation, tracking performance, and employee engagement.
- *External Advisory Councils*: External advisory councils can serve as a valuable mechanism for advancing the company's agenda by providing additional perspectives on ESG issues.

Source: Eapen, S., 2 August 2017, BSR Sustainable Business Network and Consultancy, Blog Post

www.bsr.org/en/blog/how-to-build-effective-sustainability-governance-structures

Retrieved: 28 February 2024

6.6.2 The Evolving Role of Chief Sustainability Officers

The authors report that the CSO role is transforming rapidly and dramatically. Historically, CSOs have functioned as stealth PR executives, telling appealing stories about sustainability initiatives to stakeholders while risking greenwashing. The role had virtually no involvement in setting company strategy or communicating it to shareholders; those responsibilities fell to the CEO, the CFO, and the head of investor relations. As advised in this book, some CSOs have moved away from a role centered on messaging and are spearheading the proper integration of material ESG issues into corporate strategy.

The authors state that CSOs must:

- Have a deep understanding of the company's value-creation process
- Actively engage in business transformation
- Ensure that resources are effectively distributed across the organization
- Collaborate with other C-suite executives, such as the chief financial officer, general counsel, and those responsible for ethics, risk and compliance, and
- Engage with investors to help ensure the market accurately values a company's sustainability efforts.

Four changes to the CSO role are recommended:

- Involvement in strategy and capital allocation
- More focused on and realistic about stakeholder interactions
- More fulsome engagement with investors, and
- Supported with sufficient resources and expertise throughout the organization, especially from the board and senior leadership team.

The article suggests two factors are driving this transition:

- Investors and executives increasingly recognize that sustainability is a significant factor in a company's financial performance, and
- There is political backlash in the United States against large-scale investors who incorporate ESG into their decision-making processes.

The article's authors note that there has been little substantive discussion of professionalizing and formalizing the CSO role. Instead, there have been many naively vague, incoherent, and grandiose suggestions that the CSO should take responsibility for all stakeholder interactions, innovation, and organizational culture.

A 2018 study by co-author Taylor found significant barriers to the professionalization of the CSO role, including:

- Enormous inconsistency across sectors
- A lack of role clarity, and
- Insufficient access to power and resources.

However, recent interviews of 29 CSOs across several industries and countries, plus discussions with 31 investors, found a noticeable shift in the authority and focus of the sustainability function. One shift is that CSOs are more involved in meetings with investors embracing this change. Another shift is a more central role for CSOs in companies undergoing a significant business transformation in response to existential challenges, e.g. those operating in controversial sectors and those that historically mismanage their negative external impacts.

Source: Eccles, R.G., and Taylor, A., July-August 2023, Harvard Business Review

hbr.org/2023/07/the-evolving-role-of-chief-sustainability-officers

Retrieved: 9 July 2023

6.6.3 Verizon Task Force on Climate-Related Financial Disclosures Report

The standing committees of Verizon's board of directors are responsible for monitoring the risks and opportunities related to environmental sustainability strategy and the transition to a low-carbon economy. Several directors have experience with climate-related issues, including renewable energy, network resilience, technological solutions, and emissions management.

Verizon created an ESG Center of Excellence within the operating company to implement an expanded internal control framework and facilitate compliance with climate change-related laws and regulations.

It has also established several cross-functional management councils composed of executive leadership team members that assess risks and opportunities. These councils meet regularly to address critical matters, including progress toward climate-related goals and integrating sustainability considerations into their overall strategy and business operations. The councils use a year-round planning and execution process that unites strategy development, financial planning and budgeting, talent management, and execution. These efforts are helped by scorecards to track KPIs, especially their net-zero goal.

Source: Verizon, 9 September 2021

www.verizon.com/about/sites/default/files/Verizon-2021-TCFD-Report.pdf

Retrieved: 11 November 2023

6.6.4 GE HealthCare 2022 Sustainability Report

GE HealthCare (GEHC) is embedding ESG considerations and principles into the core of its business and management systems with policy oversight by the board of directors. At the same time, the company's leadership focuses on sustainability efforts at multiple levels.

Consistent with this book's SMS model, GEHC's enterprise stewardship program (ESP) proactively identifies, assesses, and responds to risks and opportunities that could impact the company's business and operations. It has an executive-level ESG program leader who partners with the enterprise risk management (ERM) leader to manage the ESG practice and its initiatives supported by subject matter experts.

The ESG and ERM leaders co-chair the ESP Committee. The ESP Committee, composed of representatives across the company, oversees the operations of the

ESP and includes representatives from environmental, health and safety, legal, finance, and ERM, as well as the Chief Corporate Marketing and Communications Officer and the Chief Diversity, Equity, and Inclusion Officer. The latter position is a recent addition, broadening and deepening the company's ability to implement the ESP's diversity, equity, and inclusion strategy.

Source: GE HealthCare

www.gehealthcare.com/-/jssmedia/gehc/us/files/about-us/sustainability/reports/ge-healthcare-sustainability-report-2022.pdf

Retrieved: 20 September 2023

6.6.5   Walmart Environmental, Social, and Governance Highlights FY2023 and Various Corporate Website Pages

Consistent with the prime directive of finance discussed in this book, Walmart's ESG program focuses on maximizing long-term value for shareholders by:

- Serving stakeholders
- Delivering value to customers
- Creating economic opportunities for associates and suppliers
- Strengthening local communities, and
- Enhancing the business and product supply chain's environmental and social sustainability.

Due to Walmart's immense scale, the ESG program's hierarchical complexity is unsurprising. The CEO is the ultimate leader of the program, which is overseen by the board of directors' committees.

Reporting to the Executive Vice President of Corporate Affairs, the Executive Vice President and CSO updates the Walmart executive leadership team on ESG matters. The CSO administers the program across and through the company. The CSO helps define the ESG agenda and provides dedicated oversight and management of global strategic goals, priorities, and initiatives through a team led by the CSO's subordinate, the Vice President of ESG.

At least annually at the board level, the CSO provides ESG agenda and progress updates to the Nominating and Governance Committee. This committee oversees the social, community, and sustainability initiatives, charitable giving, legislative affairs, and public policy engagement strategy. Other board committees are responsible for specific ESG priority issues, e.g. the Audit Committee's oversight of the ethics and compliance program, cybersecurity, and information security.

Within Walmart's operating company, two complementary bodies function in the same capacity as the SMS model's SMS oversight group. These bodies are the ESG Steering Committee and the ESG Disclosure Committee.

The ESG Steering Committee typically meets annually to focus on strategies, efforts, and works aligned with Walmart's management strategies and priority business initiatives. It is composed of leaders from functions that drive strategies on priority ESG issues, including:

- Ethics and compliance
- People
- Global public policy and government affairs
- Sustainability, including climate strategy, and
- Corporate reporting, including representatives from the Office of the Corporate Secretary, the Controller's Office, Investor Relations, and Global Audit.

In Fiscal Year 2023, the company's corporate Disclosure Committee established the subordinate ESG Disclosure Committee to:

- Formalize established practices
- Approve and maintain ESG-report and information-governance standards, and
- Supervise, review, and monitor the preparation of ESG reports and other information for publication.

The ESG Disclosure Committee's members include the following:

- Chief Disclosure Officer
- Chief Audit Executive
- Senior Vice President, Investor Relations
- Executive Vice President and CSO
- Senior Vice President, Office of the Corporate Secretary, and
- Chief Counsel for Finance and Corporate Governance.

Source: Walmart

- corporate.walmart.com/content/dam/corporate/documents/esgreport/fy2023-walmart-esg-highlights.pdf
- corporate.walmart.com/purpose/esgreport

Retrieved: 23 February 2024

7

The Sustainability Management System's Initial Needs Assessment

This chapter addresses the first part of this question from Chapter 4:

> *How will senior leadership identify, prioritize, and shortlist the organization's most pressing sustainability needs and set strategies and tactics to be achieved within its planning cycles?*

The chapter's methods define an organization's most pressing sustainability needs at the outset of *sustainability management system* (SMS) design. These same methods can be used in subsequent SMS iterations. The results from this first needs assessment will be used to:

- Further refine the SMS's scope and scale, which were initially defined using the policy development process described in Chapter 5
- Articulate, prioritize, and shortlist the organization's sustainability needs for use in developing the SMS strategies, tactics, and targets, and
- Define the SMS's capability creation and performance initiatives.

This step in the SMS development process is a major undertaking. It produces substantial amounts of detail that require significant management efforts and resources. The requisite effort may be far more than SMS aspirants and participants have previously experienced. However, it is necessary to produce the SMS's intended outcomes.

7.1 Define the Topical Scope and Organizational Scale of the SMS

It is critically important to define the scope and scale of an SMS. For example:

Sustainability Programs: A Design Guide to Achieving Financial, Social, and Environmental Performance, First Edition. William Borges and John Grosskopf.
© 2025 John Wiley & Sons, Inc. Published 2025 by John Wiley & Sons, Inc.

- At one extreme, should an all-encompassing SMS be implemented that meets all of the demanding standards of the *United Nations Sustainable Development Goals* (UNSDGs), *Global Reporting Initiative* (GRI), *International Sustainability Standards Board* (ISSB), *International Organization for Standardization* (ISO), and other recognized authorities?
- Or, at the other extreme, is it better for an organization to occasionally do less strategic projects, such as utility and waste management cost reductions?

In reality, most SMSs will fit somewhere in between these two extremes. Therefore, it is essential that organizations carefully consider their choices. Here's why. Smaller, agile, and highly committed competitors focusing on sustainability may gain competitive advantages over organizations that choose a less committed or comprehensive approach. Further, fostering stakeholder trust, an essential intangible resource, is critically important. Therefore, it should always be proactively managed.

Considering the needs and opportunities of smaller and resource-limited organizations, the occasional sustainability initiative may be a better option. Such efforts should be easy, low-cost/high-return initiatives focused on reducing risks, production and service-delivery costs, and general expenses while potentially enhancing the organization's standing within its market segment.

With this option, resource-strapped organizations may only have to put a little extra effort into improving performance on sustainability topics. This is especially true for those already working on conventional administration and operations-focused performance improvements. For these organizations, a sustainability-themed project with a substantial *return on investment* (ROI) could be an easy inclusion in its next round of performance improvements.

7.2 The First Step in Defining an SMS's Strategic Goals, Tactical Objectives and Targets, and Initiatives

From an operational planning perspective, an SMS consists of:

- Initiating the sustainability program and setting its governing policies
- Defining, prioritizing, and shortlisting its organizational sustainability needs
- Defining strategic sustainability goals for the entire organization
- Setting tactical sustainability objectives and targets for specific units, functions, and departments, and
- Assigning specific sustainability capability creation and performance improvement projects and other initiatives to units, functions, departments, and individuals throughout the organization.

If an organization commissioned or conducted a sustainability materiality assessment as a part of its go-decision and policy development activities described in Chapter 5, that information will be an invaluable contribution to the initial SMS needs assessment. With or without a materiality assessment, a cost-effective way to start the SMS planning process is to define an organization's most pressing needs using the universally recognized *strength-weakness-opportunities-threats* (SWOT) analysis method. A fundamental advantage of this method is that it can be easily employed using available internal personnel resources, i.e. the organization's leaders and staff who thoroughly know the details of its administration and operations resources, processes, and activities. For broader considerations of sustainability needs, select external stakeholders can be invited to participate.

7.3   The Use of Outside Experts to Conduct an SMS Needs Assessment

As discussed throughout the book, organizations can retain consultants and purchase various assessment tools to perform all or parts of an SMS needs assessment. Such assessments include:

- Materiality assessments
- Stakeholder surveys
- Regulatory and industry standards compliance reviews
- Energy and greenhouse gas audits
- Waste generation inventories, and
- Life cycle assessments.

In far too many cases, these efforts are costly. Additionally, the contracted consultants will enhance their skills, but the organization's leaders may fail to learn why an SMS and the intricacies involved in its operation are needed. Such insights are required to create and manage an effective SMS to address the organization's most pressing needs.

This does not suggest that consultants and other external resources should never be used. They can provide valuable assistance for those developing, implementing, managing, or improving an SMS. The important consideration is knowing when to obtain outside help. Further, if outside help is to be procured, it is even more important to know what kinds will be needed during the various developmental phases. Therefore, it may be preferable to contract outside help when the organization clearly understands its most pressing sustainability needs. Once leaders clearly understand these sustainability needs, outside guidance can be beneficial.

7.4   An Approach to an Initial SWOT Analysis

Although other more sophisticated and complex methods are available to initially define an organization's most pressing sustainability needs, SWOT analyses are highly effective due to their ease of use, low cost, and reliance on internal knowledge and expertise.

A significant challenge for those unfamiliar with sustainability concepts is deciding which factors relevant to their organization should be analyzed. This is a critical step in defining the SMS's scope and scale. The general sustainability concerns an organization needs to consider are mentioned in Figure 2.3. Figures 7.1–7.4 provide additional detail. The topics addressed in the figures are typical:

- SMS administration concerns
- Internal operations concerns
- Upstream and downstream environmental concerns, and
- Social responsibility concerns.

Remember that despite the dozens of sustainability topics shown in the figures, they are only the tip of the metaphorical iceberg. Further, the issues in the four figures must be winnowed down to a manageable few at the outset of the needs assessment to focus on an organization's specific activities. Again, if a sustainability materiality assessment has been conducted in the development of sustainability policies, its findings can be invaluable in completing this task.

Due to its importance to the SMS's scope, design, and eventual efficacy, the initial SWOT analysis is a fairly substantial effort. It cannot be completed in an afternoon. Therefore, adequate time and resources must be set aside to produce relevant, actionable information.

7.5   Who Conducts the SWOT Analysis?

As described in Chapter 6, the SMS oversight group and the *sustainability specialty teams* (SSTs) are responsible for the various SWOT analyses of the winnowed set of sustainability topics derived from Figures 7.1–7.4. In addition, other personnel throughout the organization with expertise or interest in specific topics should also assist in this effort on an as-needed basis. This broad and deep organizational participation in defining the most pressing sustainability needs will help promote and further integrate the shared governance concept into the SMS and contribute to the sustainability program's overall buy-in.

The SMS oversight group, through the SMS champion, will assign sustainability topics to specific SSTs for analysis. During this activity, the SMS champion will

- Leadership & work force engagement & accountability

- Change management planning & execution

- SMS design, implementation, management, & improvement

- Tangible & intangible risk management

- Collaborative development of & compliance with global, national, regional, local, & industry-specific life-cycle-focused standards, targets, benchmarks & performance measurements

- Sustainability policies, strategies, tactics, & targets to improve organizational performance & create new capabilities

- Information system design, implementation, & management

- Transparency through ongoing, consistent, comparable, & verified sustainability performance sensing involving data capture, analysis, & internal & external reporting focused on:
  - Budgeting & financial control
  - Managerial accounting
  - Environmental full-cost accounting
  - ROI & related analyses
  - Financial & sustainability materiality assessments
  - Financial & regulatory compliance reporting

- Audit findings & other performance metrics issues, especially regarding the resolution of non-compliances

- Sustainability education, instruction, & training

- Equity-holder & other stakeholder engagement & collaboration

Figure 7.1 SMS Administration Concerns.
Source: W. Borges.

- Facility design, construction, commissioning, retrofitting, decommissioning, & land use issues, including *Leadership in Energy & Environmental Design* (LEED) & related standards
- Life cycle-focused value-chain management
- Sustainable product & service research & development
- Life cycle assessment of existing products & services
- Sustainable procurement
- General gaseous & specific greenhouse gas emissions obviation & mitigation
- Energy sourcing & efficiency management
- Water management – treatment, regulatory compliance & conservation
- Solid waste management
- Hazardous material management
- Hazardous waste management
- Worker health & safety management
- Greenwash-free marketing & public relations activities

Figure 7.2 Internal Operations Concerns.
Source: W. Borges.

- Stakeholder regulatory & industry standards compliance

- Periodic supplier & B2B-customer environmental stewardship & social responsibility assessments

- Natural resource management & biodiversity

- Resource extraction activities in the upstream value chain

- Direct greenhouse gas & other air emissions management upstream & downstream in the value chain

- Indirect greenhouse gas emissions reduction through energy source & activities management

- Water conservation & quality management

- Solid waste management

- Hazardous materials & waste management

- End-of-design-life *reduce-reuse-recycle* disposition of products & associated materials

Figure 7.3   Upstream and Downstream Environmental Concerns.
Source: W. Borges.

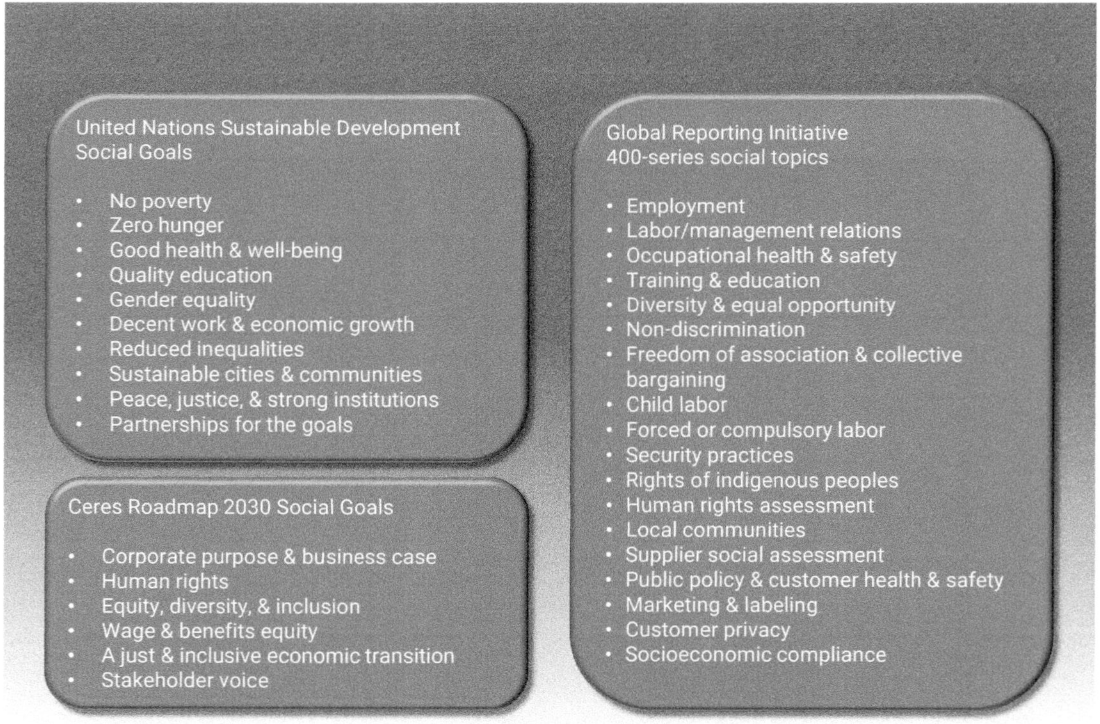

Figure 7.4 Social Responsibility Concerns.
Source: W. Borges.

assign other experts and interested personnel to help the SSTs, as appropriate. These assignments should be stated minimally in terms of:

Who will do what, by when, and how success will be determined.

Each organization has formal processes and informal conventions for assigning personnel to special projects. Follow those processes and conventions to avoid conflicts with executives, managers, and supervisors. Further, avoid repeatedly assigning the people who are often tapped for such *ad hoc* assignments. Be creative. Find up-and-comers who have special interests in specific sustainability topics. Recruit – or draft – them under the legitimate guise of professional development. The result will enhance general SMS buy-in and support, and expand the organization's personnel asset base for future sustainability projects and initiatives.

7.6 Preparing SSTs for Their SWOT Analyses

Before setting to work, SSTs need to become familiar with their respective topics. Remember, they will become the organization's *de facto* subject matter experts if they are not already. With the Internet overflowing with sustainability information, becoming familiar with basic concepts and general best practices should not be too difficult. A starting place to find topical sustainability information is Appendix Section C.7. In the mid and long terms, personnel can also gain expertise from the learning resources listed throughout Appendix C.

During their familiarization activities, each SST must become acquainted with its assigned sustainability topic's current best management practices and future state of the art. Benchmarking competitors and others in and outside an industry sector is a valuable best practice. This knowledge will help SSTs define gaps between the quality of an organization's current performance versus its potential performance.

7.7 The SWOT Analysis Process

The *internal dimensions* of a SWOT analysis shown in Figure 7.5 refer to the activities an organization controls directly. The *external dimensions* refer to its operating environment, over which it has little or no control.

Each SST will answer the following questions regarding their assigned sustainability topic(s):

- *What specific internal sustainability strengths does the organization have? Which of them needs to be built up or enhanced?*

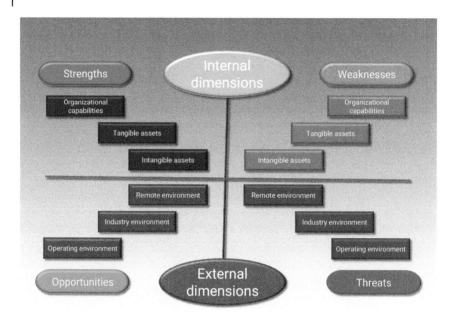

Figure 7.5 SWOT Analysis Dimensions.
Source: W. Borges.

- *What specific internal sustainability weaknesses does the organization have? Which of them needs to be reduced or eliminated?*
- *What specific internal and external sustainability opportunities does the organization have? Which of them needs to be expanded or exploited?*
- *What specific external sustainability threats does the organization face? Which of them needs to be avoided or minimized?*

Once these questions are answered for each sustainability topic, the SSTs will prioritize and summarize their findings in a concise report to the SMS champion, which will be used in the definition of the organization's most pressing sustainability needs.

7.8 Some Sustainability Ideas are Better Than Others: Prioritizing SWOT Findings

There are dozens of sustainability topics shown in Figures 7.1–7.4. When these topics are assessed using methods, such as SWOT analysis, an SMS's scope can explode in breadth and depth far beyond an organization's resource opportunities and limits. That's why the results of any needs assessment must be prioritized and then shortlisted. This basic concept is summed up succinctly by Advanced Micro Device's Tim Mohin in his 2012 book *Changing Business from the Inside Out: A Treehugger's Guide to Working in Corporations*:

> *All ideas are good ideas until you have to pay for them.*

Consistent with this sentiment, prioritization and winnowing techniques are integrated into needs assessment methods to distinguish significant sustainability issues from less important ones. The prioritization criteria can take many forms, including:

- Quick-win versus long-term *opportunities for improvement* (OFIs)
- Low-cost/high-benefit OFIs versus high-cost ones
- Risk reductions
- Administration expense and operations cost savings
- Revenue enhancements through innovation, and
- Improved transparency to enhance competitive standing.

A best practice is to save less critical issues, i.e. the ones that will not be acted on in the near future, on a *parking lot* list for future consideration. After all, current minor concerns may become significant once the major ones are resolved. Additionally, future sustainability issues will need to be considered along with these minor concerns in subsequent SMS planning cycles. Once collated, the old and new issues must be prioritized and shortlisted again.

Figure 7.6 shows a proven method for prioritizing an organization's sustainability needs. The underlying concept is the *failure modes and effects analysis* (FMEA) method, a conventional actuarial and *continuous-improvement* (CI) risk

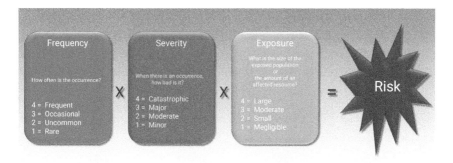

Figure 7.6 Subjective Failure Modes and Effects Analysis for Adverse Factors. Source: W. Borges.

assessment technique. FMEA is used by thousands of organizations worldwide in insurance underwriting, statistical process control, six sigma, lean manufacturing, and total quality management.

An essential attribute of this FMEA adaptation is its simplicity; it requires little quantitative input data, especially those that are hard and expensive to gather and analyze. That is because it uses readily available group consensus knowledge. Yet, despite the shortcomings of group opinions versus objective data, the method enables decision-makers to make clear distinctions between the significance of issues to establish priority rankings.

Conventional FMEA methods are focused on identifying negative efficiency factors, i.e. *failures to achieve objectives with the least cost, effort, and risk.* Negative efficiency factors in value chains that FMEA methods are used to assess include, but are hardly limited to:

- Regulatory, legal, and other sanctions, including organizational resource diversions associated with avoidable issues, such as greenwashing
- All manner of wastes, e.g. energy, solid, hazardous, liquid, gaseous, financial, personnel, time, and effort
- Mistakes
- Rework
- Delays
- Accidents
- Design flaws
- Product defects
- Cost overruns
- Internal and external health and safety risks, including those associated with defective products
- Adverse environmental impacts, and
- Social harm.

> *The most dangerous kind of waste is the waste we do not recognize…*
>
> **Shigeo Shingo**

With its exclusive focus on process failures and their effects, the conventional FMEA method works perfectly in ranking negative factors associated with a SWOT analysis's *weakness* and *threat* elements. However, the conventional method cannot characterize the positive aspects of a SWOT analysis's *strength* and *opportunity* elements. Despite this, the underlying ideas in FMEAs can be altered to rank beneficial OFIs. When FMEA's negative focus is stripped away, the method merely considers:

- The frequency of a change's occurrence,
- The exposure – i.e. size – of a set of things affected by that change, and
- The magnitude of the change's effect on that set.

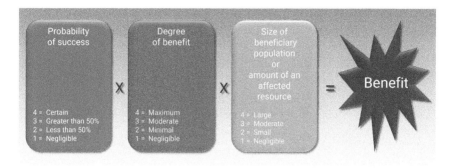

Figure 7.7   Subjective FMEA-Style Analysis for Beneficial Factors.
Source: W. Borges.

To make these factors applicable in predicting the potential benefits of *strength* and *opportunity* OFIs, they can be expressed as:

- Probability of success
- Size of the beneficiary population, and
- Degree of benefit.

This novel approach, shown in Figure 7.7, has successfully identified and ranked positive OFIs at healthcare delivery institutions in the implementation of overarching CI management systems and performance improvement processes. Of course, this same approach is helpful in any industry sector or organization.

Together, these two methods rank negative sustainability needs to be resolved and positive ones to be exploited. So, once the SWOT analysis questions in Section 7.7 are answered, each SST will score their *strength, weakness, opportunity,* and *threat* findings using the evaluation factors and formulas shown in Figures 7.6 and 7.7. The most important results will score above a statistical cut-off point set by the SMS champion. Although the universal *Pareto Principle* typically uses an 80th percentile cut-off point, it is sometimes too restrictive. (See Pareto Principle entry in the Glossary.) In such cases, many important analytical results fail to rise to a level of significance. A 50th percentile or another cut-off point might work better. SMS oversight groups need to determine what works best for them.

7.9 Aggregating, Reviewing, and Approving the SWOT Analysis Results

As noted in Section 7.7, with their work completed, each SST will deliver a report to the SMS Champion itemizing the prioritized findings from their SWOT analyses with justifications for each. In turn, the SMS champion will collate the SSTs' findings in priority order and report them to the SMS oversight group.

The SMS oversight group will review the collated and prioritized list of SWOT findings, especially in light of any previous materiality or other assessment findings. During the review, the list will be revised by deleting some findings and adding other sustainability needs consistent with senior leadership's high-level priorities. These are priorities that can lead to:

- Significant reduction or elimination of risks
- Low-cost quick-win performance improvements
- High ROI and breakthrough performance improvements, and
- High visibility – and greenwash-free – corporate image benefits.

The SMS oversight group will sort the SWOT results into five conventional management-themed categories. These five categories are the *foundation blocks, pillars*, or *elements* of the SMS, just like the ones many organizations use in enterprise-wide strategic plans:

- SMS administration
- Risk management
- Expense and cost reduction
- Revenue opportunities through innovation, and
- Competitive advantages through transparency.

> *Simplicity is the ultimate sophistication.*
>
> **Leonardo da Vinci**

As described in Chapter 6, the SMS oversight group will present its revised and sorted list to the CEO for review, amendment, and approval. This is a milestone step in the SMS model; it defines the topical scope of the sustainability program and its SMS. Once the CEO approves it, the list becomes the SMS's OFIs.

Sorted under the five conventional management categories listed above, it is easy to understand how high-ranking sustainability OFIs can be integrated into the organization's broader enterprise-wide strategic, tactical, and operational planning processes. That integration is essential to the SMS's efficacy and success. It cannot be omitted from the system's design. The lack or insufficient integration of sustainability OFIs into the overall organizational planning and control functions is an all too common sustainability program flaw.

7.10   The Next SMS Planning Steps

Chapters 8, 9, and 10 describe the SMS's remaining operations-level *planning* steps of the *Plan-Do-Check-Act* (PDCA) improvement cycle.

- Chapter 8 describes methods for using the sorted OFIs to define the SMS's strategic goals.
- Chapter 9 describes a planning process for expanding general strategic goals into a list of unit-, function-, and department-specific tactical objectives and targets.
- Lastly, Chapter 10 details how the prioritized tactical objectives and targets are operationalized via sustainability projects and other initiatives to be completed by the units, functions, and departments.
- It is important to note that Chapter 10 prescribes the SMS's operations-level *doing, checking,* and *acting* steps to complete a current quarterly PDCA cycle and start the next one.

7.11 Chapter Takeaways

- Needs assessments in the SMS development process are major undertakings. They produce substantial amounts of detail that require significant work efforts and resources. These efforts and resources may be far more than many SMS aspirants and participants have previously experienced. However, this detail is necessary to produce the SMS's intended outcomes.
- The shared governance concepts and methods prescribed in this chapter refine the scope and scale of an SMS by involving leaders and staff throughout the organization. These measures employ a conventional needs assessment method, SWOT analysis, and a novel adaptation of a common prioritization method, FMEA. The results of these needs assessment efforts are used in the:
  - Identification, prioritization, and shortlisting of the organization's most pressing sustainability needs by the SSTs
  - Review, amendment, and sorting of an aggregated list of the SSTs' most pressing sustainability needs by the SMS oversight group, and
  - Review, amendment, and approval of the most-pressing sustainability needs by the CEO to create the SMS's list of OFIs, thereby defining the topical scope of the sustainability program and its SMS.

7.12 Further Reading

The following case studies describe how Walmart and GE HealthCare define their most pressing sustainability needs, which are consistent with the prescriptions in this book.

In keeping with the Walmart program's business focus noted in Section 6.6.5, the company prioritizes *environmental, social, and governance* (ESG) topics with the most significant potential to create shared value. These topics are highly relevant to the business, stakeholders, and the company's ability to make a difference.

The first needs assessment was conducted in 2015. The most recent one, the *2021 ESG Priority Assessment*, obtained and evaluated stakeholder perspectives on over 50 topics. Walmart uses the term *ESG priority* in the same way other organizations refer to their material ESG issues.

The ESG Priority Assessment findings were prioritized and sorted into four themes: *opportunity, sustainability, community*, and *ethics and integrity*. Corporate leaders, the ESG Steering Committee, the ESG Working Group, and external stakeholders reviewed and validated the results. The findings were shared with key leaders across the organization to inform the program's aspirations, strategies, tactics, metrics, engagement approaches, and disclosures. Priorities are updated annually using information obtained from ongoing stakeholder dialogues and data gathering.

Source: Walmart

- corporate.walmart.com/content/dam/corporate/documents/esgreport/fy2023-walmart-esg-highlights.pdf
- corporate.walmart.com/purpose/esgreport

Retrieved: 23 February 2024

7.12.2 GE HealthCare 2022 Sustainability Report

Involving more than 100 internal and external stakeholders, the company's *enterprise stewardship program's* (ESP) initial needs assessment in 2021 was a materiality assessment conducted by a third party to identify the top impacts, risks, and opportunities across a wide range of ESG topics. The assessment output was a materiality matrix divided into five priority categories:

- Expanding healthcare access for underserved populations
- Promoting diversity, equity, and inclusion across the enterprise
- Mitigating climate impact and improving business performance resilience
- Advancing the circular economy and environmental design, and
- Protecting patient data and cybersecurity.

It is important to note that the term materiality, as used in this report, refers to the environmental and social impacts of the company's strategy and operations.

It does not have the same meaning used in accounting standards or under United States federal securities laws.

The company's preexisting environmental management system (EMS), formatted and certified in conformance with ISO 14001 standards, is an integral component of the ESP. The EMS identifies and documents activities, products, and services that interact with an aspect of the environment, risking or resulting in an environmental impact. A screening matrix identifies and prioritizes the most significant and impactful activities, products, and services. Detailed risk assessments further evaluate substantial risks and effects to identify specific actions to avoid or mitigate issues and optimize opportunities. These actions are recorded for implementation in four categories: *climate change, biodiversity, resource conservation*, and *pollution prevention and management.*

Source: GE HealthCare

www.gehealthcare.com/-/jssmedia/gehc/us/files/about-us/sustainability/reports/ge-healthcare-sustainability-report-2022.pdf

Retrieved: 20 September 2023

8

Define the Sustainability Management System's Strategic Goals

This chapter addresses this question from Chapter 4:

> *How will senior leadership – with the active participation of key internal and external stakeholders – define and cascade through the organization-specific sustainability goals, objectives, procedures, and accountabilities, including returns on investment (ROI) and other performance requirements?*

Chapter 7 examined using the *strengths-weaknesses-opportunities-threats* (SWOT) *analysis* method to define an organization's current sustainability advantages and disadvantages. Further, it described processes based on the *failure modes and effects analysis* (FMEAs) concept to prioritize and shortlist those advantages and disadvantages, thereby creating an initial list of *opportunities for improvement* (OFI). With the production of the initial OFIs list, the next step is to define the *sustainability management system's* (SMS) high-level strategic goals.

Goal setting has traditionally been based on past performance. This practice has tended to perpetuate the sins of the past.

Joseph M. Juran

Strategic goals based on the sustainability policy and the highest ranking OFIs provide an organization with broad themes for its future long-term sustainability initiatives. In contrast, the shorter-term tactical objectives and targets derived from the policy and strategies discussed in Chapter 9 provide greater specificity in setting qualitative and quantitative performance requirements for designated organization units, functions, and departments.

Figure 8.1 illustrates the cascading relationships between the following sustainability program elements.

- *Sustainability Policy*: The formal statement by an organization's top management declaring its overall intention to become more sustainable (Chapter 5).

Sustainability Programs: A Design Guide to Achieving Financial, Social, and Environmental Performance, First Edition. William Borges and John Grosskopf.
© 2025 John Wiley & Sons, Inc. Published 2025 by John Wiley & Sons, Inc.

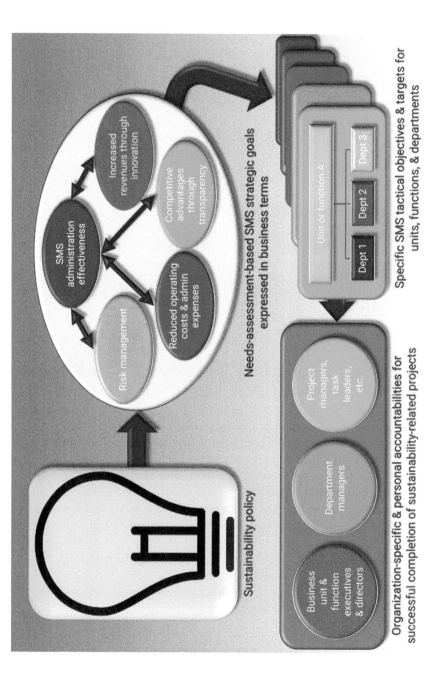

Sustainability policy

Needs-assessment-based SMS strategic goals expressed in business terms

- Increased revenues through innovation
- SMS administration effectiveness
- Competitive advantages through transparency
- Risk management
- Reduced operating costs & admin expenses

Specific SMS tactical objectives & targets for units, functions, & departments

Unit or function A

Dept 1 | Dept 2 | Dept 3

Organization-specific & personal accountabilities for successful completion of sustainability-related projects

- Business unit & function executives & directors
- Department managers
- Project managers, task leaders, etc.

Figure 8.1 Cascading Relationships Between Policy, Strategies, Tactics, and Initiatives.
Source: W. Borges.

- *Strategic Goals*: Based on needs assessments, the SMS's organization-wide strategic goals define the most critical opportunities the organization focuses on to become more sustainable (this chapter).
- *Tactical Objectives and Targets*: Once strategic goals are defined, the SMS sets accountable unit-, function-, and department-specific tactical objectives and targets to achieve the strategic goals (Chapter 9).
- *Capability Creation and Performance Improvement Initiatives and projects*: Lastly, capability creation and performance improvement initiative accountabilities are planned and completed, usually in the form of projects, to measurably achieve the tactical objectives and targets and, in turn, the strategic goals and sustainability policy (Chapter 10).

> *Improvement usually means doing something that we have never done before…*
>
> **Shigeo Shingo**

8.1 Essential Concepts in the Cascading Process

It is essential to distinguish between strategic goals, tactical objectives and targets, and unit, function, and department-level initiatives. This is because formal expectations for change are often erroneously called strategies, irrespective of an aspiration's functional degree of generality or specificity within a management system. Simply put, not all aspirations are strategic.

In *continuous improvement* (CI) management systems, the cascading relationships between policies, strategies, tactics, and initiatives must be controlled at each stage to ensure success. In contrast, Figure 8.2 shows that in *management-by-objectives* (MBOs) systems, strategic and other requirements are too frequently tossed metaphorically *over the wall* by the organization's leadership to unit, function, and department subordinates for *catch-as-catch-can* implementation. In the case of sustainability, new strategic requirements directly compete with and draw resources away from previously mandated general management accountabilities. Worse, this is often done without considering the requirements' priorities or how and when they must be implemented within the organization's resource opportunities and limits.

Without a controlled linkage process from policy development through to implementation actions, strategic goals risk remaining nebulous hopes, dreams, and sometimes institutional delusions rather than measured successes. Repeating what Stanford University professor of organizational behavior Jeffrey Pfeffer said in his 1998 book *The Human Equation*:

> *More important than having a strategy is the ability to implement it.*

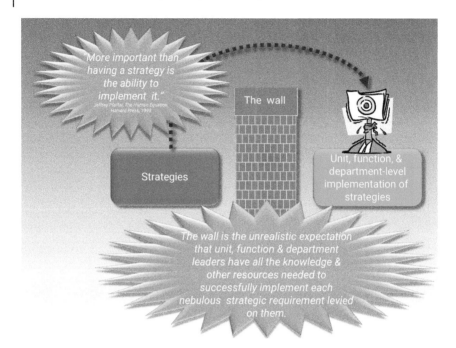

Figure 8.2   The Failure of MBO Systems.
Source: W. Borges.

8.1.1   Strategic Goals

A strategic goal is *a financial or non-financial high-concept statement of what the whole organization will achieve during a specified period* by:

- Creating new capabilities or
- Resolving significant poor-performance issues.

Strategic goals in this SMS model are derived from the sustainability policy and the prioritized list of OFIs produced during the initial and subsequent periodic sustainability needs assessments. As noted in Chapter 6, the SMS oversight group is responsible for drafting strategic goals. Once drafted, they are submitted to the CEO for review, amendment, and approval.

Strategic goals, such as time-sensitive waste, water, and energy-use reductions, may be expressed quantitatively. However, for many strategic goals, this is not appropriate. Strategic goals are big ideas, the details of which are specified when defining individual tactical objectives and targets for organization units, functions, and departments. These big ideas are subsequently detailed during the development of tactical objectives and targets using the *SMART* format that describes *specific, measurable, achievable, relevant,* and *time-constrained* objectives and targets.

8.1.2 Tactical Objectives

Tactical objectives *are more specific qualitative prescriptions preferably with quantitative targets for time-limited actions that contribute to achieving one or more broad strategic goals.* Accountabilities for a tactical objective may be levied on one, several, or all organization units, functions, and departments.

Tactical objectives that meet the *SMART* criteria are typically achieved via detailed project plans. As described below, the SMS oversight group led by the SMS champion, the sustainability specialty teams (SSTs), and the accountable unit, function, and department-level project managers, must closely monitor and collaboratively manage these project plans to measurably achieve the assigned tactical objective(s).

8.1.3 Organization-level Unit, Function, and Department-level Initiatives and Projects

Organization-level unit, function, or department-level initiative and project plans *specify how an assigned tactical objective and target will be achieved.* The SSTs assist accountable managers in preparing the plans. Initiative plans for simple to achieve objectives and targets – aka *Point-Kaizen events, tiger- and red-team interventions, small-tests-of-change,* and *quick-wins* – can be expressed in such basic terms as:

Who will do what, by when, and how success will be measured.

However, complex objectives and targets require detailed project plans featuring *work breakdown structures, schedules, budgets,* and *personnel assignments.* Chapter 10 and Appendix E provide essential rudiments of project design and management.

Because implementation plans are created at an organization's unit, function, and department levels, their authors are able to consider and balance day-to-day realities in project design and implementation efforts. These realities include their daily work activities and *key performance indicators* (KPIs) on dashboards, balanced scorecards, and other periodic financial and managerial accounting reports. Once sustainability accountabilities are assigned to organization units, functions, departments, and even individuals, the associated tactical objectives and targets must be added to their respective KPIs to establish equivalency with other accountabilities.

8.2 Converting Prioritized OFIs into Strategic Sustainability Goals

Chapter 7 explored a process to define high-priority sustainability OFIs in these five planning elements:

- *SMS Administration*: OFIs for improving the SMS itself
- *Risk Management*: Sustainability-related regulatory compliance, industry standard, contractual requirement, and legal sanction conformance OFIs
- *Operating Cost and Administrative Expense Reductions*: Efficiency OFIs to resolve sustainability-related poor-performance issues with the least cost, effort, and risk
- *Revenue Opportunities through Innovation*: Sustainability-themed revenue growth and generation OFIs, and
- *Competitive Advantages through Transparency*: Organizational transparency, competitive advantage, and other tangible and intangible benefits OFIs.

This last point, *transparency*, is often a confusing buzzword for leaders when first creating an SMS. Why is it so important? Figure 8.3 provides a few of the concept's potential concrete and intangible benefits.

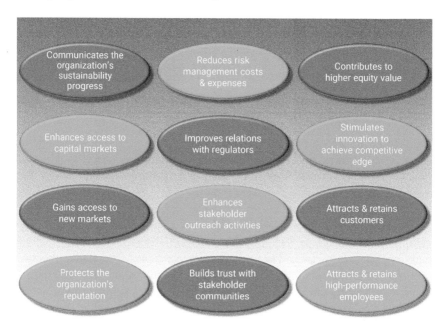

Figure 8.3 Examples of Sustainability's Potential Transparency Benefits.
Source: W. Borges.

The next step is to convert the OFIs into draft high-level strategic goals. The SMS champion and the organization's strategic planning director – plus any other key personnel they deem appropriate – are responsible for converting the OFIs into a draft slate of strategic goals. Why include the strategic planning director in this task when the SMS has a designated champion? The SMS must be integrated into the broader organization's planning and management processes to be most effective. Who knows how to do this better than the strategic planning director? However, in smaller organizations lacking a strategic planning director, other members of the SMS oversight group should be recruited to assist with this task.

Converting OFIs into strategic goals is a relatively simple process compared to the SWOT analysis. The *strengths, weaknesses, opportunities,* and *threats* boil down into these two planning categories:

- *Advantages*: The OFIs responding to strengths and opportunities and
- *Disadvantages*: The OFIs responding to weaknesses and threats.

The advantage OFIs are converted into active-voice statements declaring those:

- *Strengths* that need to be enhanced and
- *Opportunities* that need to be exploited.

Here are two examples of strategic goals with and without quantitative mandates. The first goal establishes an organization's time-limited climate commitment for not only its internal operations, but its entire value chain. The second goal seeks to exploit an organization's sustainability and training and development expertise to create an external public relations advantage through corporate transparency concepts.

- Example Strategic Goal 1: *The organization will achieve carbon neutrality within its direct operations during the next 5 years and throughout its entire value chain during the next 10 years.*
- Example Strategic Goal 2: *The organization will become a recognized industry leader in sustainability management by developing experts within its ranks and files.*

The disadvantage OFIs are likewise converted into active-voice statements declaring which:

- *Weaknesses* need to be eliminated or reduced and
- *Threats* need to be avoided or minimized.

Here are two active-voice examples of strategic goals that seek to resolve hypothetical internal performance issues and external risks with and without quantitative mandates. The first is intended to reduce currently excessive expense burdens, liabilities, and risks associated with waste management. The second addresses a need to avoid future *environmental, health, and safety* (EHS) and other regulatory sanctions.

- Strategic Goal 3: *The organization will reduce waste disposal expenses by 25% within two years by aggressively applying reduce, reuse, and recycle principles.*
- Strategic Goal 4: *The organization will become its industry's benchmark for achieving all EHS regulatory mandates with the least cost, effort, and risk.*

Subsequent tactical objectives and targets will provide specificity for general strategic goals lacking time constraints and high-level goals.

It is impossible to produce superior performance unless you do something different from the majority.

Sir John Templeton

Once completed, the draft slate of strategic goals is submitted by the SMS champion to the SMS oversight group for review and amendment. Once the goal reviews and revisions are complete, they are forwarded to the CEO for final review, amendment, and approval.

8.3   Performance Breakthrough Goals

The preceding process can produce fairly straightforward strategic goals. These are generally adequate during the SMS's first few years and iterative cycles. However, leaders of highly effective SMSs should strive to define *breakthrough performance* goals, often called *stretch goals.*

It is a rare group of senior leaders – in this case, the SMS oversight group – that does not have collective epiphanies while reviewing mundane draft strategic goals. During such reviews, they often realize they have been too narrow in their focus and meek in their aspirations; therefore, they must think bigger and bolder. To avoid limited or timid goal development, these leaders should critically examine the draft sustainability goals from their high-level leadership perspectives to ensure the organization's overall achievement of its mission, vision, values, and especially financial performance. Then, it is simple for them to strengthen weak goals and add or delete others. One assessment approach to produce stronger and breakthrough goals is to answer the following questions for each prospective strategic goal:

- *How do the organization's draft goals compare to the benchmarks set by its competitors and leaders in other industries?*
- *Irrespective of benchmarks, what does ideal performance look like for each goal?*
- *What obstructions have kept the organization from already achieving ideal performance?*

- *What are the root causes of these obstructions as defined through brutally honest Ishikawa or similar cause-and-effect analyses?*
- *Lastly, what are the bold strategic goals needed to eliminate these root causes?*

After these amendment considerations, the SMS oversight group submits the slate of draft strategic goals to the CEO for review, amendment, and acceptance. The result is the SMS's sustainability strategies.

Chapter 9 uses these initial strategic goals to start building the critical links – i.e. the specific tactical objectives and targets – between general intentions and highly detailed day-to-day work activities.

8.4 Frequency of Defining and Updating Strategic Goals

Every well-run organization has some kind of strategic planning cycle. Conventional three to five-year strategic plans are the norm at many conservative organizations, while shorter rapid-cycle plans are produced at some of the more progressive ones. Irrespective of the timeframe, an SMS's planning cycle should be designed to fit within the overall organizational planning cycle. Within that overall cycle, the SMS's strategic goals must be updated at least annually as new information on sustainability opportunities and marketplace conditions becomes available.

8.5 Future Needs Assessments and Strategy Development

This initial strategic goal definition process relies on the SWOT analysis method described in Chapter 7. Subsequent updates may even continue using this method. Here are the results an organization can expect from its SMS after its first few iterations:

- SMS processes are operating more efficiently and produce measurable benefits
- Leaders and personnel have become more proficient in producing measurable results
- The SMS's initial and subsequently updated strategic goals have been largely achieved, and
- The organization gains insights and confidence from its early positive results.

Despite these accomplishments, though, new opportunities and risks inevitably arise. SMS performance can begin to level off, leading to complacency and underachievement, endangering long-term SMS effectiveness. At this point, more exacting needs assessment methods are needed to counter this negative dynamic and assure progressively effective SMS results. Plus, increasingly sophisticated financial, environmental, and social impact assessment methods are required to capture, analyze, and report SMS performance data.

Such assessment methods include limited-scope regulatory and industry standards audits and stakeholder surveys. Beyond audits and surveys, sustainability materiality assessments can be invaluable when greater detail is needed to identify broad ranges of stakeholder concerns. Life cycle assessments (LCAs) are instrumental at the value-chain level when greater detail is required to identify environmental stewardship and social responsibility OFIs at specific points in these closed-loop systems. Various LCA considerations are discussed in *Appendix F*. These methods can be conducted internally or by outside contractors depending on organizational needs and available resources.

8.6 Chapter Takeaways

- This chapter explained the critical cascading relationships between an SMS's policy, strategies, tactics, targets, projects, and initiatives.
- It first focused on a work process using the OFIs from Chapter 7 to define the SMS's set of high-level strategic goals.
- Further, it described a way to define additional breakthrough performance goals from an executive perspective.
- Then, it previewed how this initial strategy definition process can become more sophisticated and refined during subsequent SMS improvement cycles using such needs assessment methods as audits, surveys, sustainability materiality assessments, and LCAs.
- Moving on from this strategy development process, it previewed how Chapter 9 prescribes methods for defining and distributing specific tactical objectives and targets throughout an organization's units, functions, and departments.

8.7 Further Reading

The following case study describes how Johnson and Johnson uses its needs assessments to produce its sustainability program's strategic goals. The goals to be achieved by 2025 are listed.

8.7.1 Johnson & Johnson Health for Humanity 2025 Goals

Johnson & Johnson (JNJ) produces *Priority Topics Assessments*, i.e. sustainability-focused materiality assessments, to guide long-term policy and strategic goal processes. Like Walmart and GE HealthCare, JNJ retains a qualified specialty consulting firm to conduct these needs assessments every two or three years, which is appropriate considering the immense sizes of these companies. Using specialty consulting firms helps these global-scale companies efficiently obtain uniform and impartial information on environmental, social, and governance (ESG) topics.

The needs assessment findings are sorted and prioritized into four ESG focus areas supporting 11 of the 17 *United Nations Sustainable Development Goals* (SDGs) for *global health, social justice, environmental stewardship,* and *responsible business practices*. The four ESG focus areas are:

- Champion global health equity
- Empower employees
- Advance environmental health, and
- Lead with accountability and innovation.

It is important to note that on the surface, JNJ's ESG focus areas do not explicitly address the wide range of operational EHS risk management considerations. This is because the company operates with a well-established *ISO 14001*-certified *management system* (MS) that addresses environmental issues. At the same time, health and safety matters are managed under the *ISO 45001 Occupational Health and Safety Standard*. It should also be noted that the company encourages but does not require its operating units to achieve ISO 45001 certification. These two EHS standards and their MSs are integral to the overarching ESG program.

With mature and effective EHS MSs in place, the ESG program office led and overseen by the Chief Sustainability Officer and the Regulatory Compliance and Sustainability Committee of the Board of Directors, produces long-term sustainability strategies and goals in alignment with the SDGs.

Strategic goals are defined once the needs assessment findings are sorted and prioritized. The current goals for the four ESG focus areas to be achieved by 2025 are:

- Champion global health equity
  - Global access plans
  - Access to HIV treatment
  - Access to tuberculosis treatment
  - Developing tuberculosis treatment
  - Access to schizophrenia treatment
  - Support frontline health workers
  - Reduce the burden of obstetric fistula
  - Preventive viral vaccine capabilities
  - Vaccination monitoring platform our goals
  - Healthcare for women

- Empower employees
  - Women in management
  - Ethnic/racial diversity in management
  - Black/African Americans in management
  - Healthiest workforce score
  - Women in stem
- Advance environmental health
  - Renewable electricity
  - Carbon neutrality for company operations
  - Scope 3 emissions reductions
- Lead with accountability and innovation
  - Supplier sustainability
  - Global supplier diversity and inclusion
  - Partnerships for good

The next step in JNJ's ESG program is using these goals to define KPIs for business units, functions, and departments.

Source: Johnson & Johnson

- www.jnj.com/about-jnj/policies-and-positions/priority-topics-assessment
- www.jnj.com/about-jnj/policies-and-positions
- www.jnj.com/health-for-humanity-goals-2025
- www.jnj.com/about-jnj/policies-and-positions/our-position-on -environmental-health-and-safety-management

Retrieved: 9 April 2024

9

Define the Sustainability Management System's Tactical Objectives and Targets

This chapter continues answering this question from Chapter 4:

> *How will senior leadership – with the active participation of key internal and external stakeholders – define and cascade through the organization-specific sustainability goals, objectives, procedures, and accountabilities, including returns on investment and other performance requirements?*

It does so with a high-level planning method essential to operationalizing sustainability. This method uses the sustainability management system's (SMSs) broad strategic goals to craft specific tactical objectives and targets to be achieved by discrete organization units, functions, and departments. These objectives and targets are the links between strategic sustainability intentions and capability creation and performance improvement projects and initiatives.

9.1   A Deeper Dive into the Linkage Concept

Although discussed in Chapter 8 and illustrated in Figure 8.2, it still may not be clear as to:

> *Why is it necessary to spend so much time and effort linking strategic intentions to day-to-day activities? Aren't the well-paid unit, function, and department leaders supposed to be doing this already?*

For many organizations attempting sustainability programs, this is where the implementation process often fails to operationalize lofty intentions. In far too many organizations, the procedural linkages between strategic intentions and

Sustainability Programs: A Design Guide to Achieving Financial, Social, and Environmental Performance,
First Edition. William Borges and John Grosskopf.
© 2025 John Wiley & Sons, Inc. Published 2025 by John Wiley & Sons, Inc.

day-to-day work activities – especially *ad hoc* projects and other intitiatives – are weak or nonexistent. This last point is at the heart of this book, i.e.:

> *Systematically and systemically operationalize sustainability to avoid program underperformance or failure.*

So, rather than metaphorically tossing strategic goals *over the wall* as shown back in Figure 8.2 and hoping for the best, this chapter delves deeply into a highly structured process. It is designed to parse out what each organization unit, function, or department must do to contribute to the achievement of individual strategic sustainability goals. Failure to complete this activity risks program underperformance, such as the intermittent implementation of a few unrelated and ineffective environmental and social-benefit projects. Although well-intentioned, such efforts cannot transform the organization into a sustainable one.

As noted earlier, the reason the *management-by-objectives* (MBO) concept can work well at large, old-line-industry companies is because they have well-developed policies, procedures, and other resources, including many *masters of business administration* (MBA) degree holders and other graduate-level management professionals whose jobs involve translating strategic goals into measurable results. However, smaller organizations – especially technology-based enterprises such as architecture/engineering/construction, healthcare, specialty manufacturing, logistics, and information technology firms – typically have only a few of these management professionals. As a result, organization unit, function, and department leaders typically lack the expertise to translate an organization's broad strategic intentions into specific measurable improvements.

What's the fix? Remember back to the earlier discussion around Stanford Professor Jeffrey Pfeffer's idea:

> *More important than having a strategy is the ability to implement it.*

And, of course, there's always the classic adage:

> *Every system is perfectly designed to produce the results it gets.*

This adage has a corollary:

> *To change results, the system must be changed.*

These ideas provide insights into why MBO, initially conceived by Dr. Peter Drucker, does not work as well as it could in many smaller organizations, especially technically oriented ones. These insights hint that instead of large numbers of MBAs, organizations might benefit from the design, implementation, and active management of highly structured linkage processes that guide non-MBA decision-makers in making the right decisions at the right times. This is where continuous improvement (CI) management systems are most effective.

Introducing such processes and systems for a sustainability program requires formal instruction with periodic follow-up and reinforcement for the SMS's various participants shown in Figure 6.1. Following basic adult education principles, such instruction should be based on adaptations of the materials in this book and practical *learn-by-doing* experiential activities. It ingrains the necessary CI and sustainability concepts into the knowledge and skill sets of the SMS leaders and participants.

As noted in earlier chapters, there are two primary components in this linkage:

- Tactical qualitative objectives and quantitative targets focused on specific organization units, functions, and departments, and
- Capability creation and performance improvement projects and other initiatives at the organization unit, function, and department levels to achieve those objectives and targets.

This chapter examines the first linkage component, transforming strategic goals into tactical objectives and targets. Organization unit, function, and department-specific projects and other initiatives that achieve tactical objectives and targets will be discussed in Chapter 10.

9.2 A Process for Converting Strategic Goals into Tactical Objectives and Targets

The process described here for developing detailed tactical objectives and targets from broad strategic goals is conceptually simple. It uses an essential analytical tool, the *matrix,* like the example in Figure 9.1. Matrices list two sets of interrelated factors, one on the vertical axis and the other on the horizontal axis. The intersection points on matrices draw attention to the potential for significant relationships between two discrete factors within the larger sets.

As shown in Figure 9.1, the following two matrix factors are used to identify the organization units, functions, and departments that will participate in achieving specific strategic goals.

- The organization's units, functions, and departments are listed across the horizontal axis.
- Strategic goals defined in Chapter 8 are sorted and listed down the vertical axis in the SMS's five organizational management categories:
  - SMS administration
  - Risk management

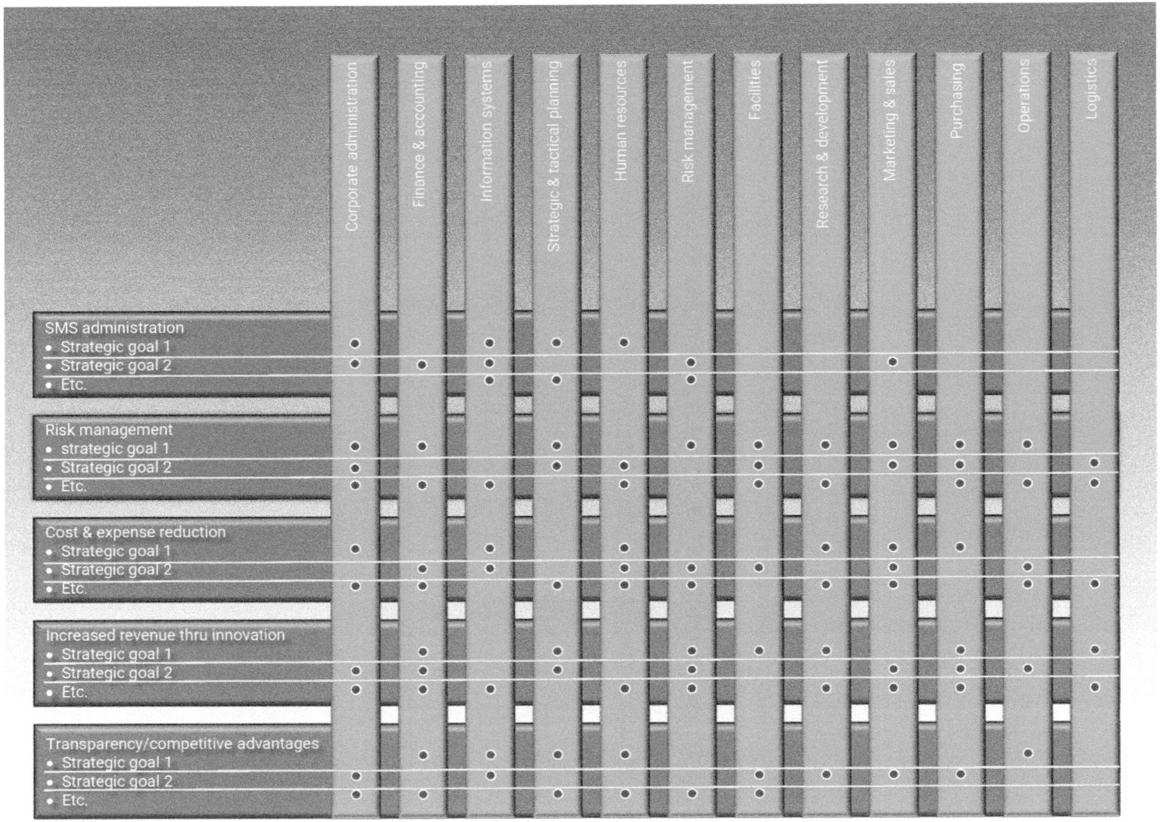

SMS administration
- Strategic goal 1
- Strategic goal 2
- Etc.

Risk management
- strategic goal 1
- Strategic goal 2
- Etc.

Cost & expense reduction
- Strategic goal 1
- Strategic goal 2
- Etc.

Increased revenue thru innovation
- Strategic goal 1
- Strategic goal 2
- Etc.

Transparency/competitive advantages
- Strategic goal 1
- Strategic goal 2
- Etc.

Corporate administration
Finance & accounting
Information systems
Strategic & tactical planning
Human resources
Risk management
Facilities
Research & development
Marketing & sales
Purchasing
Operations
Logistics

Figure 9.1 Tactical Objectives and Targets Assignment Matrix.
Source: W. Borges.

- Cost and expense reductions
- Revenue opportunities through innovation, and
- Competitive advantages through transparency.

The draft tactical objectives and targets produced by the matrix analysis will be collaboratively refined by:

- The SMS champion
- The SMS oversight group
- The (SSTs)
- Organization unit, function, and department leaders, and
- And, ultimately, the *chief executive officer* (CEO).

Three types of information are produced by the matrix analysis:

- Prescriptions for capability creation and performance improvement actions *(Note: Not every intersection on the matrix warrants a prescription for action.)*
- Specification of the organization units, functions, and departments that must complete the prescribed capability creation and performance improvement actions, and
- Definition of the actions in terms of the specific qualitative objectives and quantitative targets by which success will be determined. *(Note: Although each tactic should have a quantitative target, this is not always possible.)*

This information should be drafted into terse, active-voice tactics using this format:

- *Who,* as in which organization unit, function, or department
- Is going to do *what* in terms of specific qualitative and quantified actions
- By *when,* and
- *How* will tactical objective and target success be unambiguously verified?

Further, the *SMART* concept is used to evaluate and strengthen the potential efficacy of a draft tactic, i.e. is it *specific, measurable, achievable, relevant,* and *time-constrained*? Failure to draft and assess tactics considering the SMART concept risks setting ineffectual and unimplementable expectations, especially if the answers to the *when* and *how* questions lack specificity.

9.3 SMS Oversight Group, Sustainability Specialty Teams, and Other Personnel Roles

The SMS oversight group is responsible for defining the SMS's tactics. As noted in Chapter 6, because of the multidisciplinary work efforts typically required to define actionable objectives and targets, the SSTs will assist. Further, the SMS oversight group will draw on additional personnel resources throughout the organization whenever needed.

9.4 Examples of Tactics

The examples of tactics discussed below are derived from the following single strategic sustainability goal, which was presented in Chapter 8:

> *The organization will achieve carbon neutrality within its direct operations during the next 5 years and throughout its entire value chain during the next 10 years.*

Note that this goal is focused on modifying an organization's direct operations and its value chain, which are management concerns driven by environmental stewardship and social responsibility needs. The emphasis on direct operations and the value chain helps integrate the derived tactical objectives and targets into an organization's overarching planning, budgetary, and control processes as risk management and cost and expense initiatives. As noted earlier, expressing SMS tactics in management terms helps make them intellectually and emotionally palatable to resistant leaders and others who are not yet convinced of sustainability's business proposition.

The tactics examples below show how a single strategic sustainability goal can be applied across an organization in distinctly different ways from unit to unit, function to function, and department to department. It is important to note that these are only some tactics that could be derived from this one goal. Plus, in keeping with a sub-theme of this book, i.e. *one size does not fit all*, each organization must define its tactical objectives and targets aligned with its unique strategic goals.

- Example Tactic 1: *By the end of the first fiscal quarter of 2025 (Q1/25), the facilities department, in collaboration with the SMS champion, will complete an organization-wide energy-use audit by contracting with an industry-leading energy/greenhouse gas auditing firm. Energy-reduction opportunities in the audit report will be provided to the SMS oversight group via the SMS champion for executive action within the SMS. Additional requirements for the audit are to define the organization's baseline greenhouse gas footprint and identify immediate, midterm, and long-term opportunities for improvement (OFI) to become carbon neutral.*
- Example Tactic 2: *Acting on the energy-use audit findings, the SMS champion will direct the design and implementation of an organization-wide*

energy-management and carbon-reduction program. The SMS oversight group will review, amend, and approve the program. Development of the program will start in the third quarter of 2025 (Q3/25) and be implemented in the fourth quarter of 2025 (Q4/25).

- Example Tactic 3: *Consistent with the program's requirements created by Example Tactic 2, the information systems department will design a focused project plan to reduce electricity use in data and communications systems throughout the organization. The plan will be implemented by the end of the second fiscal quarter of 2026 (Q2/26).*
- Example Tactic 4: *By the end of the fourth fiscal quarter of 2025 (Q4/25), the purchasing department, in collaboration with the facilities department, will identify competitively priced, on-grid, sustainable-source electricity suppliers. Within these criteria, it will then enter into purchase agreements, enabling the maximum possible progress toward achieving the strategic goal.*

9.5 Refining the List of Tactical Objectives and Targets

As noted earlier, the processes that define specific tactics and targets to implement broad strategic goals in particular organization units, functions, and departments can produce overwhelming numbers of new projects and other requirements. It is here that recognizing an organization's resource opportunities and limits is critical. Therefore, winnowing those actions to a manageable number consistent with the organization's resource capabilities is necessary. In some organizations, the number of initiatives assigned to a particular unit, function, or department might be as many as five. In contrast, it might be just one or none in others. Determining that number is a management responsibility of the SMS oversight group.

Once the SMS oversight group drafts candidate tactics with the assistance of SSTs and other personnel, they must be prioritized. This is where the beneficial approach to *failure mode effects analysis* (FMEA) shown in Figure 7.7 can be helpful again. The SMS oversight group can again use this method to draft a shortlist from the top-scoring tactics. Please note that Figure 7.6's negative FMEA method will not likely be needed. This is because, by the time tactics have been defined, the underlying negative issues from the earlier needs assessment have been transformed into beneficial OFIs that are the bases for the SMS strategies and tactics.

Once drafted, the SMS oversight group further evaluates and revises these shortlisted tactics in consideration of factors such as these:

- Compliance requirements
- Returns on investment

- Marketplace opportunities
- Reputation considerations
- Quarterly sequencing and scheduling opportunities and constraints
- Availability of resources
- Capacities of organization units, functions, and departments to successfully meet tactical objectives and targets, especially considering competing accountabilities, and
- Equitable distribution of accountabilities among organization units, functions, and departments.

In some organizations, senior leaders in the SMS oversight group may be comfortable with a consensus-based evaluation process involving internal experts with the relevant expertise to assist with the shortlist evaluations and revisions. In others, senior leaders may prefer a more objective evaluation criteria scoring process designed in-house.

During this evaluation, the SMS oversight group must consult with leaders in the affected organization units, functions, and departments to determine if draft tactics can be accomplished using available *labor, equipment, workspace, vendor support, materials,* and *budgets.* If not, the draft tactic must be amended to state that additional resources will be needed. Further, if the draft tactic requires other resources, the SMS oversight group must secure budgetary commitments.

Regarding labor resources, consideration must be made regarding whether the need can be met through in-house development, interdepartmental transfers, additional hires, or third-party contracting. Negotiations between the SMS oversight group and the organization unit, function, and department leaders are implicit in such evaluations and should be a recognized and welcomed part of the collaboration process. As a part of this process, the SMS oversight group will define the criteria by which projects and other actions are funded or rejected.

After this evaluation, the SMS oversight group will amend the resulting list of tactical objectives and targets by:

- Combining and culling redundancies, and
- Removing any draft tactics that do not rise to a level of significance, as determined by the SMS oversight group, after further consideration.

Keeping with the concept that *one size does not fit all*, each organization must determine its significance thresholds for this process.

As with the strategic goal process described in Chapter 8, the SMS oversight group will submit the final list of tactics to the CEO for review, amendment, and approval. Once approved, the SMS oversight group will assign the tactics as organization unit, function, and department-level accountabilities to achieve the strategic sustainability goals.

It is critically important to maintain rejected lower-priority candidate tactics on a separate list. These lower-priority tactics will be considered again in later SMS cycles when the current tactics have been achieved. At that time, many lower-priority tactics may move up on the list for implementation. However, with time, some low-level draft tactics may become irrelevant and must be removed from the list.

The next chapter, Chapter 10, prescribes a process for designing and successfully completing specific time-limited organization unit, function, and department projects and other initiatives to achieve tactical objective and target accountabilities.

9.6 Chapter Takeaways

- This chapter describes how the SMS's strategic sustainability goals are used to draft tactical objectives and targets to be achieved by discrete organization units, functions, and departments.
- The definition of tactics is a high-level organizational planning activity is the responsiblity of the SMS oversight group. In contrast, implementing tactics is an organization unit, function, and department activity discussed in Chapter 10.
- The SMS oversight group drafts new tactical objectives and revises them periodically, typically quarterly. The SSTs and other personnel may be called upon to assist.
- The examples of tactics in this chapter show how individual strategic goals can be implemented through multiple organizational initiatives in various organization units, functions, and departments.
- Because this methodology can produce multiple draft tactics for each strategic goal, they must be winnowed into a manageable number consistent with an organization's resource opportunities and limits. The winnowing analyses conducted by the SMS oversight group focus on such factors as:
  - Compliance requirements
  - Returns on investment
  - Marketplace opportunities
  - Reputation considerations
  - Quarterly sequencing and scheduling opportunities and constraints
  - Availability of resources
  - Capabilities of organization units, function, and departments to successfully meet objectives and targets, especially considering competing accountabilities, and
  - Equitable distribution of accountabilities among organization units, functions, and departments.

- The SMS oversight group submits the draft list of tactical objectives and targets to the CEO for review, amendment, and approval.
- The approved list of tactical objectives and targets will be used to assign capability creation and performance improvement accountabilities to specific organization units, functions, and departments to achieve sustainability projects and other initiatives.

9.7   Further Reading

The following case study describes how Boeing develops strategic goals into tactical objectives and targets. Further, it emphasizes the importance of management disciplines to achieve beneficial outcomes. A technical criticism, though, is the objectives and targets lack the SMART concept's specificity.

9.7.1   The Boeing Company Sustainable Aerospace Together: 2023 Sustainability Report

This case study illustrates the adage, *every system is perfectly designed to get the results it gets.* The company's technical, operational, labor, legal, and other difficulties that were widely reported in 2024 arose because, despite its intentions and performance standards, there were failings in the management disciplines required to effectively implement them. The situation is consistent with the idea presented earlier in this book, … *a poorly designed and managed organizational structure enables undesirable behaviors leading to underperformance or failure.*

On the surface, though, Boeing's program structure appears sound. Intermediate tactics and targets link overarching strategies to organization units, functions, and departments. This is promising in that the program is attempting to hardwire the sustainability program's aspirations to produce measurable performance improvements. The company's recent difficulties provide important lessons in organizational management that, if applied, can lead to more certain achievement of the goals and objectives stated below.

The following examples from its 2023 sustainability report illustrate how Boeing began the cascading process in the current iteration of its sustainability program to refine high-level strategies into more specific tactical objectives and targets.

- Employee Safety and Well-Being
    - Strategic Goal: Value human life and well-being above all else and act accordingly; strive to prevent all workplace injuries.

  - Tactical Objectives and Targets:
    - Top quartile recordable injury rate (among sector benchmarked performance).
    - ≥90% believe their manager supports their well-being.
- Global Aerospace Safety
  - Strategic Goal: Drive aerospace safety to prevent accidents, injury, or loss of life with our Boeing culture and actions rooted in safety.
  - Tactical Objective and Target: Drive aerospace safety via global aerospace safety initiatives to maintain the downward trend of the worldwide commercial jet fleet's 10-year moving average fatal accident rate.
- Equity, Diversity, and Inclusion
  - Strategic Goal: Address representation gaps and strengthen equity, diversity, and inclusion so all team members feel supported and inspired to reach their full potential.
  - Tactical Objective and Target: Increase representation of women globally and underrepresented racial/ethnic minorities in the United States.
- Sustainable Operations
  - Strategic Goals:
    - Maintain a net-zero future for Boeing operations through conservation and renewable energy.
    - Partner with the supply chain for responsible business practices.
  - Tactical Objectives and Targets:
    - Achieve a 55% absolute reduction in Scope 1 and Scope 2 GHG emissions from the 2017 base year.
    - Maintain net-zero emissions for Scope 1 and Scope 2.
    - Achieve 100% renewable electricity.
    - Work with suppliers to increase GHG reporting and proactively address risks driven by climate change-driven risks.
- Innovation and Clean Tech Strategy
  - Strategic Goal: Enable the transition to carbon-neutral aerospace through investments and partnerships for fleet efficiency improvements, sustainable aviation fuel (SAF), and future platform technologies.
  - Tactical Objectives and Targets:
    - Current and future commercial airplanes will be 100% SAF compatible.
    - Build and certify our first zero-emission, electric, autonomous aircraft via the Wisk Aero subsidiary.
    - Support the commercial aviation industry's ambition to achieve net-zero carbon emissions for global civil aviation operations by 2050.
- Community Engagement
  - Strategic Goal: Build better, more equitable communities through corporate investments, employee engagement programs, and advocacy efforts.

– Tactical Objective and Target: Expand opportunities for more than 12.5 million youth, veterans, veterans' families, and underserved individuals across communities around the world.

Source: The Boeing Company, 21 December 2023
www.boeing.com/sustainability/annual-report
Retrieved: 1 April 2024

10

Capability Creation and Performance Improvement Initiatives

This chapter completes the answer to this high-level planning question from Chapter 4.

> *How will senior leadership – with the active participation of key internal and external stakeholders – define and cascade through the organization-specific sustainability goals, objectives, procedures, and accountabilities, including returns on investment and other performance requirements?*

Additionally, it answers these *sustainability management system* (SMS) operations-level project design and performance questions from Chapter 4.

> *How will single- and multiple-quarter sustainability capability creation and performance improvement projects be designed and completed at all business unit, function, and department levels?*

> *How will sustainability capability creation and performance improvement project performance be monitored and reported to confirm that goals, objectives, and targets are on track to successful completion?*

Next, it answers these operations-level questions about project evaluations:

> *How will projects lasting more than one quarter be evaluated?*

> *How will end-of-project reviews be conducted to evaluate the degree of project success?*

> *When projects fall short, how will corrective actions be defined and implemented?*

> *How will project successes be celebrated and rewarded?*

Lastly, it answers these high-level SMS improvement questions from Chapter 4.

Sustainability Programs: A Design Guide to Achieving Financial, Social, and Environmental Performance, First Edition. William Borges and John Grosskopf.
© 2025 John Wiley & Sons, Inc. Published 2025 by John Wiley & Sons, Inc.

How will lessons learned from projects – especially the definition of new best practices – be documented, communicated, and successfully implemented?

How will the list of the organization's most pressing sustainability needs be updated and re-prioritized quarterly and annually?

How will lessons learned regarding the efficiency of the SMS – i.e. the ability to achieve sustainability aspirations with the least cost, effort, and risk – be used to improve it?

How will the next iteration of the PDCA-based SMS be initiated for the next fiscal quarter, fiscal year, or other period?

Everything in the book to this point describes preparations for an organization to become more sustainable. However, operationalizing sustainability requires specific needs-based actions that produce measurable beneficial results. Using the tactical objectives and targets developed in Chapter 9, this chapter focuses on the assignment, design, tracking, and effectiveness evaluation of capability creation and performance improvement projects and other initiatives. The process concentrates on the operations level of the *do, check,* and *act* phases of the SMS's PDCA cycle.

Keep in mind that not all sustainability initiatives are projects. Some can be accomplished within hours, if not minutes, through small Point-Kaizen events, tiger- and red-team interventions, small-tests-of-change, and quick-win improvements. As such, they typically do not require formal project plans and their associated management controls. Organizations already using these approaches merely need to adapt existing processes to include sustainability-focused small-change initiatives. Others not currently using them should explore and implement rapid-response methods best suited to their needs.

> *All improvement happens project by project and in no other way.*
>
> **Joseph M. Juran**

This chapter concludes by discussing the importance and utility of lessons learned and how the PDCA cycle is restarted during the SMS's *act* phase. In doing so, it describes a process to improve capability creation and performance improvement outcomes.

With this chapter's prescribed methods, the detailed SMS process map illustrated in Figure 2.4 progresses from the *plan* phase through to the *act* phase, completing an entire PDCA cycle. Figure 10.1's Gantt chart schedule shows the chronological sequence of the SMS model's quarterly PDCA cycle activities described in this chapter. Organizations will, of course, customize the quarterly process to meet their administrative needs, which may differ from the figure.

Quarterly SMS activities	Weeks														
	Week 0	Week 1	Week 2	Week 3	Week 4	Week 5	Week 6	Week 7	Week 8	Week 9	Week 10	Week 11	Week 12	Week 13	Week +1
Preparations for the quarterly SMS meeting by the SMS champion and the SMS oversight group with the assistance of support functions.	■														
Quarterly SMS meeting: assignment of new capability creation and performance improvement accountabilities by the SMS oversight group.		■													
Projects and other short-term initiatives designed, reviewed, amended and initiated by accountable unit, function and department leaders with the assistance of assigned SSTs and the SMS champion with the support of the SMS oversight group.			■												
Project tasks and other short-term initiatives completed by accountable unit, function and department project and initiative teams with the advisory support of the assigned SSTs.			■	■	■	■	■	■	■	■	■	■			
Weekly project progress and short-term initiative completion reports to the SMS champion by the accountable project and initiative managers.		✳	✳	✳	✳	✳	✳	✳	✳	✳	✳	✳	✳		
Bi-weekly programmatic progress reports to the SMS oversight group by the SMS champion.			✳		✳		✳		✳		✳		✳		
End-of-project completion reports and multi-quarter-progress reports by accountable project and initiative managers to the SMS champion.													✳		
Programmatic end-of-quarter report prepared for the SMS oversight group by the SMS champion.													✳		
Quarterly SMS meeting: SMS oversight group evaluations of project and other initiative performance with follow-up actions regarding incentive awards, lessons-learned, and new opportunities for improvement.														■	
New SMS administrative improvement accountabilities assigned by the SMS oversight group.														■	
The SMS's quarterly PDCA cycle formally restarted at the quarterly SMS meeting.														■	
Quarterly SMS meeting findings recorded, archived and distributed to SMS stakeholders by support functions.														■	

Figure 10.1   Quarterly Schedule of SMS PDCA Cycle Activities.
Source: W. Borges.

10.1   Accountability Assignments

As described in Chapter 9, the SMS oversight group will assign capability creation and performance improvement accountabilities to specific organization units, functions, and departments, as well as to their leaders and key personnel. Thirteen-week fiscal quarters are the most common frequency. To achieve the best outcomes, the assignment process should be conducted in an open, collaborative meeting between the SMS oversight group, the associated SSTs, and the affected leaders and key personnel from the organization units, functions, and departments. This is both the beginning and – as will be seen later in the chapter – the end of quarterly SMS improvement cycles.

Collaboration between these SMS participants is crucial in achieving buy-in and support for the proposed sustainability projects. It also deepens the mutual understanding of the purposes of the SMS, the importance of its initiatives, and the ways they will be accomplished. Concurrently, the SMS oversight group learns the challenges organization unit, function, and department leaders face in implementing sustainability initiatives along with their general management accountabilities. This prepares the SMS oversight group to better support these leaders.

The exclusive purpose of the inaugural SMS meeting is to assign the first round of sustainability projects. After that, though, the purpose of subsequent quarterly meetings will be expanded to review, assess, and improve project and other initiative performance, along with the assignment of new accountabilities. The project and intiative review process is described later in this chapter. These quarterly meetings are essential to ensure that the SMS's policies, strategies, tactics, and targets are effectively achieved throughout the organization. These achievements elevate sustainability needs up to par with the organization's other general management priorities.

> **A Note About Communications Problems**
>
> "Lack of communication" is an all too common complaint. Whenever this problem arises, the root cause is typically a deficiency in the work process. While there can be other reasons for insufficient communication, a root cause analysis will most likely point to omitted steps in the work process. In addition to missing critical steps, there are probably undefined conditional decision points. Conditional decision points in a stepwise process provide essential "if this situation, then go to this step" considerations that lead logically to the correct next steps. In these common situations, misnamed communications problems require contemporary – e.g., Lean, Six Sigma, or Total Quality Management (TQM) – process improvement solutions. See *Continuous Improvement Methodologies* in the Glossary.

As mentioned, the quarterly meetings provide a critically important opportunity to involve the organization unit, function, and department leaders in evaluating, negotiating, and determining the following:

- Availability of resources
- Capabilities of units, functions, and departments to successfully meet objectives and targets, especially in consideration of competing general management accountabilities
- Equitable distribution of accountabilities among organization units, functions, and departments, and
- Resolution of communication problems between the SMS oversight group, SSTs, and affected organization units, functions, and departments.

Understanding the purpose of this collaborative assessment and negotiation process is essential. The hard-working and often over-stressed organization unit, function, and department leaders and key personnel may regard the quarterly meetings as opportunities to reject the assigned accountabilities. However, at this point in the process, the rejection of an accountability is a dead issue.

In response to any pushback, it must be made unequivocally clear that the *chief executive officer* (CEO) has mandated the tactical objective and target. The mandate is to resolve one or more of the organization's most pressing sustainability needs. Therefore, the issues to be resolved during quarterly meetings must be focused exclusively on how the assigned tactical objective and target will be accomplished with the least cost, effort, and risk.

From such discussions, the SMS oversight group, the associated SSTs, and the accountable leaders can begin to define and negotiate any additional resources needed to complete the assigned capability creation and performance improvement projects and initiatives.

Why is this collaborative assessment so critical? Conceptually, the role of organization leaders is simple. Their three primary responsibilities are:

- *Make their numbers*;
- *Run their part of the organization without surprises*; and
- When either of these two things is not working, *improve performance.*

Realistically, though, organization leaders' roles are incredibly complex. Sustainability accountabilities risk potentially unrealistic burdens being added to an already challenging job. That is why organization unit, function, and department leaders must be involved in assessing opportunities and constraints in collaboration with the SMS oversight group and the SSTs. These are their respective roles:

- The SMS oversight group supports the leaders by clearing actual and potential obstacles to project success, especially the need for additional resources. When additional resources are required, the SMS oversight group must support the organization unit, function, and departments in identifying and acquiring them. Of course, there will be times due to resource constraints when it is necessary to delay the initiation of a project or pause its performance. The SMS oversight group is responsible for making such decisions.

- Supporting the SMS oversight group, the role of the SSTs is to assist the organization unit, function, and department leaders in finding technically sound solutions during project planning and execution.

10.1.1 SMS Incentive Programs

A key element in assigning personal sustainability accountabilities to organization unit, function, and department leaders, along with key project team participants, is the provision of adequate incentives to complete projects successfully. By providing sufficient incentives, sustainability initiatives are elevated in importance alongside the leaders' other general management accountabilities. Failure to incentivize sustainability accountabilities is a definite way to ensure they receive inadequate attention or are ignored outright.

It can be argued that implicit threats to job security are motivations enough to complete sustainability initiatives. However, carrots rather than sticks can do a better job of introducing and institutionalizing new, conceptually unfamiliar programs, such as sustainability. The same applies to insignificant incentive opportunities. Further, while it is essential not to be stingy, it is also important to recognize that nonmonetary rewards are valuable enhancements to incentive packages.

An elemental responsibility of executive leadership in an SMS is to determine and understand personnel motivations and then design incentive programs to make it worthwhile for everyone in the organization to embrace and support sustainability enthusiastically. Incentive programs must be well designed to reward behaviors that promote and maintain sustainability efforts. Positive motivators and inducements at every level of the organization – especially at the governing board and top management levels – are highly effective in achieving desired sustainability outcomes.

10.1.2 Accountability Tracking

Once collaborative opportunities and constraints negotiations are complete and the assigned accountabilities are finalized, they must be entered into the organization's overall *management system* (MS) tracking and reporting processes as *key performance indicators* (KPIs). Organizations with contemporary *enterprise resource planning* (ERP) systems likely have modules for recording such accountabilities and the means to track and report project progress. However, smaller organizations may not have such extensive information systems. Creating a rudimentary tracking system from scratch will be necessary in these cases.

Effective tracking and reporting processes are the requisite activities of the *check* function in the PDCA cycle, regardless of an existing ERP or a scratch-built system.

10.2 Project Design

As a part of assigning an accountability, the SMS oversight group will provide the organization unit, function, or department with technical and administrative support via an appropriate SST. As the go-between for the SMS oversight group and the accountable organization function, unit, or department, the SST(s) will assist the accountable leaders in:

- Producing a viable project plan
- Communicating extraordinary project resource needs and issues back to the SMS oversight group via the SMS champion
- Assuring objective milestone progress reports are produced
- Helping define root causes and corrective actions in response to variances and nonconformances to the project plan, as well as supporting the successful completion of corrective actions, and
- Producing an objective end-of-project assessment, including articulation of lessons learned.

Organizations with contemporary performance improvement programs – e.g. Lean, Six Sigma, and TQM – will use established processes to design sustainability projects. Such organizations have performance improvement experts who should be involved in assuring SMS project design processes and formats are consistent with existing standards.

Organizations lacking such programs must create processes to design detailed project plans. Additionally, organizations new to formal project management methods may need to obtain instruction through outside resources, such as those listed in Appendix C. The following sample worksheets are provided in Appendix E for personnel needing assistance in rudimentary project planning:

- Figure E.1, Project Summary Worksheet
- Figure E.2, Task Description Worksheet
- Figure E.4, Project Schedule Worksheet, and
- Figure E.5, Project Budget Worksheet.

Preparing a budget for all projects may or may not be necessary, especially for smaller short-term performance improvement projects and initiatives using existing organizational resources. The SMS oversight group should make such decisions on a case-by-case basis. Also, the chief financial officer, in collaboration with other members of the SMS oversight group, may consider establishing a contingency budget item for small-scale sustainability projects and initiatives and an expenditure ceiling below which formal project budgeting is unnecessary.

Complex and large-scale capability creation and some performance improvement projects may have terms extending over several fiscal quarters. In such cases, projects should be designed in quarterly phases. In contrast, initiatives such as small *Point-Kaizen events, tiger-* and *red-team interventions, small tests of change,* and *quick-win* improvements can resolve problems in minutes, hours, or, at most, a day or two.

Project design is an urgent, time-constrained activity that should be completed within a week or two of an accountability being assigned. With the assistance of their responsible SSTs, the accountable leaders will forward the completed project plans to the SMS champion. In those organizations with contemporary ERPs, key elements of project plans – e.g. schedules, budgets, and milestones – must be entered into the performance tracking and reporting modules along with the assigned accountability. In organizations lacking this capability, key elements of project plans must be included in the accountability tracking and reporting system developed in-house, as noted earlier.

10.3   Requests for Additional Project Resources

Be aware that as an element of project planning, fulfilling requests for additional resources is also an urgent activity. Many project plan budgets will define the need for extra internal labor, outside expertise, space, equipment, and materials. This is especially the case for capability creation projects. In contrast, performance improvement projects can often be accomplished with existing resources. This is because currently available resources are not being used *efficiently* – i.e. they are not achieving the work process objectives with the least cost, effort, and risk – hence the need for performance improvement.

The additional resource needs defined in project plans must be forwarded to the SMS champion as formal requests, including itemized budgets using the organization's standard formats. As needed, the SMS champion will assess and revise the requests as needed with inputs from the responsible SSTs, accountable leaders, responsible departments, and other knowledgeable resources. Then, the SMS champion will communicate these requests to the SMS oversight group. With the SMS oversight group's review, amendment, and approval, the SMS champion will expedite the resource requests via the organization's established accounting and procurement procedures.

Because organizations must operate within their resource opportunities and limits, additional resource requests should be expected to be modified. It's the idea expressed by this *Rolling Stones* lyric:

> *You can't always get what you want, but if you try sometime, you'll find you get what you need.*

Resource request modifications may also require alterations to project plans. The SMS champion, SMS oversight group, and the SSTs will assist the accountable leaders in making such changes. The SMS champion and accountable leaders will then record the changes in the tracking and reporting systems.

10.4 Milestone Progress Reports

Organizations with existing enterprise-wide CI MSs may need to adjust their current project tracking and reporting processes to report SMS project progress. In organizations without a CI program, though, formal progress-reporting processes will need to be defined for the following types of documentation activities.

- *Ad hoc monitor-detect-correct (MDC) action reports*: These reports record:
  – Programmatic and in-the-moment observation of work in progress
  – Documentation and reporting of variances and nonconformances to project plans, and
  – Corrective actions to resolve variances and nonconformances to project plans.
- *Project progress reports from SSTs to the SMS champion*: The reporting frequency may vary due to the project's size, importance, and other factors, but weekly reporting is common. The reports will focus on project performance metrics and qualitative characteristics that reflect conformance to *schedule, budget,* and *performance quality* requirements. Special attention must be paid to any plan deviations, their root causes, and the efficacy of corrective actions.
- *Completion reports from rapid-response initiative leaders to the SMS champion*: These reports will describe the purposes and outcomes of actions involving *Point-Kaizen events, tiger and red-team interventions, small-tests-of-change*, and *quick-win* improvements.
- *A portfolio-level summary report of all projects prepared periodically by the SMS champion for the SMS oversight group*: These reports will emphasize key performance metrics and qualitative factors. The typical frequency is biweekly, but this may vary. The information will highlight plan deviations, root causes, and the

efficacy of corrective actions. Further, the report will itemize issues requiring action by the SMS oversight group.

These reporting processes require support from the decision support and information systems functions during and after the SMS's initial development. The more these activities can be automated, the easier it will be for SMS participants to focus on completing sustainability projects and less on administrative distractions.

10.5   Assistance in Defining Root Causes and Corrective Actions

Accountable leaders in organizations with effective CI MSs will likely have the skills to define root causes when deviations from project plans are detected during performance monitoring. They may also be adept at designing effective corrective actions to put projects back on course.

However, accountable leaders often need assistance with both tasks in organizations new to CI concepts. Leaders and staff responsible for quality assurance, organizational effectiveness, and performance improvement should be able to assist accountable leaders with root cause analyses (RCA). However, if such expertise is unavailable, it must be acquired through in-house instruction, outside assistance, or a combination. A short list of RCA learning resources is presented in Appendix C's Section C.6.2.

RCA is an assessment method used to identify the underlying causes of work process problems and other difficulties that may not be obvious or are hard to determine. Failure to identify true root causes is an all too common issue, especially for organizations new to CI methods. It is essential to eradicate the root cause of a problem once and for all instead of reacting repeatedly to its consequences, which needlessly wastes time and other resources. The most common RCA method is the *Ishikawa diagram* method, also known as a fishbone diagram, herringbone diagram, cause-and-effect diagram, or even Fishikawa.

Root causes should be defined quickly – e.g. within an hour or so of discovering a plan deviation – so effective corrective actions can be designed and implemented as soon as possible. Note that complex plan deviations may require lengthier team-scale efforts. Without a speedy root cause determination, a project may experience additional problems from schedule delays and other issues that waste resources.

Once the root cause of a project plan deviation is identified, accountable leaders may need additional help designing and implementing effective corrective actions. The responsible SSTs – with the help of the organization's specialists responsible for quality assurance, organizational effectiveness, and performance improvement – must be available to help the accountable leaders. In some cases, the SMS champion should intervene directly or assign others to do so. As with RCAs, corrective actions must be designed and implemented quickly to avoid schedule delays and their subsequent impacts.

10.6 End-of-Project and Multi-Quarter-Project Assessments

Organizations operating with enterprise-wide CI MSs may need to adjust their current monitoring standards and processes to incorporate the SMS's projects, initiatives, and other activities. In contrast, organizations lacking an overall CI MS will likely need to define and implement project monitoring standards and processes from scratch. These include:

- Single- and multi-quarter progress report and end-of-project report formats
- Formal project closure processes
- Criteria for determining project success, plus processes for their celebration
- Processes for evaluating unsuccessful or underperforming projects, as well as defining and completing corrective actions
- Processes for reporting project outcomes to the organization's chief officer leadership, enabling them to award incentives to the projects' accountable leaders and team members, and
- Standards and processes for articulating, cataloging, and disseminating lessons learned.

With the SMS's end-of-project assessment process established, the SMS champion, on behalf of the SMS oversight group, will schedule and usually lead meetings for key project stakeholders at the end of each fiscal quarter or other established performance period. Such meetings should coincide with but precede the beginning-of-quarter PDCA-cycle restart meetings when new accountabilities are assigned.

- Such meetings should include the SMS champion, SMS oversight group, and all key project participants, again using an open meeting format to encourage stakeholder engagement, involvement, and feedback to produce consensus outcomes.
- However, there may be occasions when the meeting should be closed to only the reviewers and the accountable leaders with their project teams.

The former has the advantage of providing all SMS participants with opportunities to learn from the efforts of other project teams. Whereas the latter closed meeting format has the benefit of a degree of confidentiality when corrective actions involve personnel actions or the project being discussed is organizationally sensitive.

10.6.1 Unsuccessful Project Considerations

Regarding projects with unsuccessful outcomes, it is critically important to reiterate this key change management concept:

> *Never punish people for the sins of a work process.*

When a project is unsuccessful, it is the collective responsibility of the SSTs, the SMS champion, and the SMS oversight group to determine the root causes of deficient project outcomes. This is critical in articulating lessons learned and prescribing improvements to future projects and the overall SMS. Unless an accountable leader and key project team members obstinately prevent the successful completion of a project, there is little justification for punitive action on the part of the SMS oversight group or its members. Leaders and team members who put their best efforts into a project and, yet, come up short should be applauded – and in some cases even rewarded – for uncovering a need for an SMS administration improvement, staff development need, or a new sustainability *opportunity for improvement* (OFI).

It is important to note that after decades of performance tracking in formal CI MSs, the root causes of most project failures are flawed systems and work processes. Even when project problems involve personnel, the root causes typically center on failures to provide people with adequate competency-focused instruction and other support resources.

10.6.2 End-of-Fiscal-Quarter Project Reports and Reviews

During the end-of-quarter meetings, the SMS oversight group will review and evaluate summary reports of project completion or progress with each accountable leader. These reports must be standardized and included in the SMS tracking system along with each project's other documentation. Further, this documentation must be available throughout the organization on a need-to-know basis to communicate positive and negative lessons learned. This is intended to enhance the probability of success with current and future sustainability initiatives. Lastly, such documentation helps increase the overall awareness of SMS activities and their effectiveness by internal and external stakeholders.

The following information should be included in end-of-project and multi-quarter-project progress reports:

- For short-term projects of one fiscal quarter or less, the accountable leaders will prepare a summary documenting:
  - The assigned accountability
  - A concise description of the project
  - Evidence of successful project completion, including the project's effectiveness in contributing to the achievement of tactical objectives and targets, preferably in quantitative terms, plus on-time and within-budget performance
  - Or, in the case of unsuccessful projects, the root cause and prescriptions for corrective and preventive measures for future projects
  - Lessons learned, including suggestions for next steps, and
  - Suggestions for enhancements to SMS work processes.
- For multi-quarter progress reports, the accountable leaders will prepare a similar summary documenting:
  - The assigned accountability
  - A concise description of the project with emphasis on the current fiscal quarter's completed tasks
  - Evidence of acceptable project implementation to keep on schedule and within budget while maintaining task performance quality
  - In the case of project plan deviations, a statement of the root causes, plus an evaluation of the corrective action efficacy
  - Lessons learned
  - A list of the next performance period's tasks with predictions of expected challenges requiring obviation and mitigation measures to enhance the probability of project success, and
  - Any suggestions for enhancements to the overall SMS work processes.

The SMS oversight group's end-of-fiscal quarter reviews will focus on the following:

- Objective factors – especially quantitative ones – demonstrating project success or failure
- In the cases of multi-quarter projects, the next tasks
- In the case of multi-quarter-progress reports, foreseeable challenges requiring obviation and mitigation measures
- Assessment of accountability achievement
- Any incentive awards and associated project celebrations
- Lessons learned, and
- Possible new OFIs.

10.7   Restart the Quarterly SMS Improvement Cycle

This section describes the all-important *act* phase of the SMS's quarterly PDCA cycle involving capability creation and performance improvement initiatives. Note that the separate annual review process of the overall SMS is described in Chapter 11. The quarterly *act* phase completes an iteration of the PDCA cycle and starts a new one. This is the PDCA cycle's fundamental strength, i.e. continuous system improvement.

With the completion of quarterly reviews, the SMS oversight group will:

- Modify operations-level SMS performance standards and work processes, as needed
- Assess any new OFIs discovered during the end-of-quarter reviews using the tactical objective and target development process in Chapter 9, and include in support of current strategic goals, any new high-priority tactical objectives and targets in the implementation schedule
- Delete from the annual implementation schedule any tactical objectives or targets that are no longer relevant or required to achieve specific strategic goals
- Reprioritize the annual tactical objective and target implementation schedule
- Update the list of tactical objectives and targets to be implemented next quarter along with their proposed accountabilities
- Submit next quarter's list of tactical objectives, targets, and accountabilities to the CEO for review, amendment, and approval, and
- With the CEO's approval, assign new capability creation and performance improvement accountabilities to specific units, functions, departments, and their leaders and key personnel.

In conclusion, it is necessary to emphasize that time is of the essence with all activities associated with any CI MS. Combined, the end-of-fiscal-quarter review and the restart of the quarterly SMS improvement cycle are significant work efforts. These activities have the potential for inexperienced organizations to waste time and effort, thereby risking frustration with – and even failure of – the overall sustainability program. The challenge for any organization, especially the smaller ones, is to develop and continually refine efficient overlapping work processes for the reviews and PDCA cycle restarts. This is so that everything from preparation efforts to the conclusion of the proceedings can be accomplished quickly and efficiently, ideally within a week. Further, participants must set aside adequate time to perform their responsibilities successfully.

This quarterly SMS review and PDCA cycle restart process is initially complex, however, it becomes easier after a few iterations. Because it has a broader, higher-level purpose, the annual administrative review of the overall sustainability program described in Chapter 11 should be conducted separately.

10.8   Chapter Takeaways

- In this chapter, the SMS activities progress from the *planning* phase of the PDCA cycle to the *doing, checking,* and *acting* phases, moving from preparing to change the organization to doing the work to change it.
- The processes described in this chapter are significant work efforts that must be continually refined to improve SMS performance, thereby contributing measurably to the organization's overall success in becoming more sustainable.
- The SMS champion, the SMS oversight group, and the SSTs must collaborate with the organization unit, function, and department leaders to assess opportunities and constraints associated with accountability assignments and their projects.
- The SMS oversight group supports accountable leaders by clearing potential obstacles and creating conditions for project success.
- The SSTs assist the accountable leaders in developing technically sound solutions.
- Adequate incentives are needed to motivate leaders and project team participants to enthusiastically embrace sustainability and raise the significance of SMS projects and initiatives to levels comparable to those of other general management accountabilities.
- SMS tracking and reporting processes are needed to control project progress. Organizations with existing reporting systems may need to adjust them to accommodate SMS activities. Organizations without such systems need to create them.
- Project accountabilities must be assigned, performance tracked and documented, and lessons learned made available as reference materials to drive SMS project and system-wide improvements.
- Project Design Considerations
  - Project designs and resource requests are time-sensitive activities requiring rapid action. Performance improvement, organizational effectiveness, quality assurance, and other experts should assist in these efforts when available.
  - The standard term for conventional performance improvement projects is a 13-week fiscal quarter. However, large-scale projects may extend over several quarters, requiring multiple quarterly phases.
  - Point-Kaizen event, tiger and red-team intervention, small-tests-of-change, and quick-win improvements produce immediate small-scale solutions, often within minutes, hours, or a day or two.
- Project Tracking and Reporting
  - Accountable leaders will use the MDC concept to proactively manage project performance and report progress to the SMS championvia the tracking system.

- Root causes must be defined and corrective actions must be implemented quickly to avoid project delays, further problems, and wasted resources.
- The SMS champion will prepare a biweekly portfolio-level performance report of all projects, including deviations, root causes, and corrective actions, for the SMS oversight group.
- At the end of each fiscal quarter, the SMS oversight group will review and evaluate all project progress and completion reports with each accountable leader. Discussions will focus on success and failure factors, the efficacy of corrective actions, further challenges, lessons learned, OFIs, next steps, and achievement of accountabilities.
- It is critical that accountable leaders and project team members put their best efforts into a project. However, when they provide their best efforts and come up short, they should be applauded – and sometimes even rewarded – for uncovering a need for an SMS administration improvement or a new sustainability OFI. They should never be punished for flaws in the SMS's work processes.
- End-of-quarter SMS reviews are the end and start of new quarterly iterations of a PDCA-based improvement cycle. Note that the quarterly iterations operate within annual PDCA cycles.
- The annual administrative reviews of the overall sustainability program described in Chapter 11 should be held separately from this chapter's quarterly operations-level reviews.

10.9 Further Reading

Consistent with the book's prescriptions, the two examples below illustrate complete cascading processes whereby a single strategic goal is linked through increasingly specific tactics and targets that guide the development and completion of projects and other initiatives to produce measurable outcomes.

10.9.1 Microsoft 2022 Environmental Sustainability Report

Microsoft established a *Climate Council* of senior company leaders to govern sustainability progress and priorities. The Council is analogous to the book's SMS oversight group. Focusing on *company, customer,* and *global* sustainability topics, the company cascades its people, planet, and profit aspirations into its administration and operations units to create new capabilities and improve performance. It publishes scorecards twice yearly and reviews progress quarterly.

The following examples show how one strategy to achieve *zero direct-generated waste* is linked to several discrete accountable initiatives assigned to organization units, functions, and departments. The initiatives and projects range from securing certifications and adopting new industry standards to building new end-of-design-life processing facilities and reducing and diverting wastes. Notably, the activities described below are just some of the efforts to achieve this strategy.

Strategic Goal: *Zero Waste by 2030 Across the Company's Direct Waste Footprint.* Tactical Objectives and Targets:

- Drive to zero waste operations.
- Increase reuse and recycling of servers and components.
- Eliminate single-use plastic in all primary product packaging and all IT asset packaging in data centers by 2025.
- Make fully recyclable products and packaging.

Major Projects, Initiatives, and Outcomes:

- Waste Diversion: zero waste certifications were renewed for several United States and Ireland data centers. In FY22, 12,159 metric tons of solid waste generated at the company-owned data centers and campuses were diverted from landfills and incinerators.
- Cloud Hardware Reuse and Recycling: Four new reuse and recycling Circular Centers opened during FY22 in the United States, Ireland, and Singapore. Two additional Circular Centers will be operational in the United States by 2025. These facilities enabled cloud hardware server and component reuse and recycle rates to reach 82% in FY22.
- Plastic Reduction: In FY22, single-use plastics in product packaging were reduced by more than 29%, a 4.7% decrease to 3.3% by the average weight of plastic per package.
- Device Recyclability: The company is switching to a new EU energy-related products recycling standard, EN45555, to assess product and packaging recyclability.

Source: Microsoft, 31 May 31 2023
query.prod.cms.rt.microsoft.com/cms/api/am/binary/RW15mgm
Retrieved: 18 March 2024

10.9.2 Mattel 2022 Citizenship Report

Mattel's *ESG Executive Council* is chaired by the Chairman and CEO and is composed of key senior executives. Analogous to the book's SMS oversight group, the Council defines the company's ESG strategy and goals using information obtained periodically during ESG materiality assessments. The most recent iteration of this needs assessment was conducted in 2020. The Council also evaluates and approves ESG programs and plans in support of Mattel's purpose and objectives.

Implementing the *GHG Reduction Roadmap* described below is overseen by a separate *Operations Sustainability* (OS) function. The OS function operates similarly to the book's SSTs and receives periodic updates on progress against KPIs.

As with Microsoft's program described above, the following examples show how one strategy to reduce energy consumption and *greenhouse gases* (GHGs) is linked to multiple discrete accountable initiatives assigned to organization units, functions, and departments. The initiatives and projects range from procurement changes to capital-intensive capability creation projects.

Strategic Goal: Reduce absolute Scope 1 and 2 GHG emissions by 50% by 2030 versus a 2019 baseline.

Tactical Objective: In 2022, Mattel developed an internal *GHG Reduction Roadmap* to guide energy and GHG reduction activities in five areas –

- Reduction of energy demand
- On-site solar
- Off-site solar
- Clean energy procurement, and
- Renewable energy certificate purchases.

Major Projects, Initiatives, and Outcomes:

As part of the Roadmap, Mattel developed 25 core capability creation and performance improvement initiatives to reduce emissions, including detailed schedules for solar projects. Project-specific details are not provided in the Citizenship Report. However, overall energy and CO_2 reduction results attributable to these actions from 2019 to 2022 are reported as follows:

- Reduction in Absolute Energy Consumption by Type in MWh
  - Direct Absolute Energy Consumption – 21%
  - Indirect Absolute Energy Consumption – 13%
  - Total Absolute Energy Consumption by Type – 15%
- Reduction in Absolute GHG Emissions by Scope in Metric Tons of CO_2e
  - Scope 1 Absolute GHG Emissions – 24%
  - Scope 2 Absolute GHG Emissions – 10%
  - Total Absolute GHG Emissions for Scope 1 + 2 – 12%

Source: Mattel, Inc., 18 September 2023
assets.contentstack.io/v3/assets/bltc12136c3b9f23503/blt9629cb310f0aedf5/650b6a914b35c8d0fdfffe76/Mattel_Citizenship_Report_FINAL.pdf
Retrieved: 18 March 2024

Annual Administrative Assessment and Improvement of the SMS

This chapter answers the questions in Chapter 4 regarding the administrative assessment and improvement of the overall *sustainability management system* (SMS).

> *How will lessons learned regarding the efficiency of the SMS – i.e. the ability to achieve sustainability objectives with the least cost, effort, and risk – be used to improve it?*
>
> *How will the next iteration of the PDCA-based SMS be initiated for the next fiscal quarter, fiscal year, or other period?*

The operations-level quarterly review and improvement process described in Chapter 10 is just one aspect of the PDCA cycle's *check* and *act* functions in an SMS. The other is an annual administrative assessment and improvement of an SMS's overall ability to plan, organize, control, and lead in achieving sustainability policies and goals. This annual review is one of the SMS's primary inputs to an organization's *environmental, social, and governance* (ESG) reporting activities. Remember the distinction between SMSs and ESG activities. The purpose of sustainability programs is to *manage* an organization's environmental, social, and financial concerns, while the purpose of ESG is to *report* on the management of those concerns. To support both sustainability program and ESG efforts, the purpose of the annual SMS assessment is to methodically ask:

- *Is the overall sustainability program achieving its strategic goals?*
- *If so, what is the empirical proof?*
- *If not, why not?*
- *In either case, how is the organization going to respond?*

Sustainability Programs: A Design Guide to Achieving Financial, Social, and Environmental Performance, First Edition. William Borges and John Grosskopf.
© 2025 John Wiley & Sons, Inc. Published 2025 by John Wiley & Sons, Inc.

This chapter prescribes a process to annually assess and administratively improve the overall SMS, which includes:

- An annual internal SMS assessment by an organization's upper management
- Internal audits, inspections, and assessments, and
- External audits, inspections, assessments, and regulatory actions.

Lastly, this chapter addresses the importance of an SMS's high-level *check* and *act* functions in supply chain management to avoid people, planet, and profit risks.

11.1 The Annual SMS Administrative Assessment

A high-level annual assessment is one of the most important *check* and *act* activities in an SMS. It assesses the SMS's efficacy as an internal organizational management tool leading to performance improvements. As noted, it also provides a significant amount of the information needed to produce voluntary and mandated sustainability and ESG reports.

Given the functional differences between SMS quarterly reviews and annual assessments, the annual assessment meeting should be conducted separately from the last quarterly review meeting of the year.

The SMS oversight group or other chief-officer-level bodies should conduct the annual meeting. Typically, the SMS champion chairs it with the participation of board of directors members, the *chief executive officer* (CEO), the SMS oversight group, and other responsible executives. Further, depending on the organization's size and complexity, key personnel from the organization units, functions, and departments will participate in various specialty roles.

As a formal assessment, it is conducted strictly with a standardized agenda. It is suggested that the organization's most knowledgeable and experienced support staff be assigned to participate in this review under the direction of the SMS champion. They will record the minutes of the assessment discussions in detail and summarize them with a focus on the following:

- Key findings, lessons learned, and decisions
- Follow-up assignments to improve the overall performance of the SMS, and
- Timely distribution and continued availability of minutes to internal and external stakeholders on a need-to-know basis.

The total time required for the annual assessment meeting will vary depending on an organization's size, complexity, and effort in documenting the preceding year's SMS performance. For smaller, less complex organizations, a few hours may suffice. However, a full day or more may be needed when there are many issues. In contrast, larger, more complex organizations may spend several days. Preparations and follow-up activities may require several days for some participants before and after the meeting. Determining factors are the accessibility and readiness of the requisite performance information. Unsurprisingly, one of the SMS's earliest high-priority OFIs will be the automated capture, analysis, and reporting of SMS performance information.

The following is a best practice format for documenting any annual management system (MS) assessment that has evolved over three decades of use in the private sector and increasingly in the public sector. It includes key elements of the *International Organization for Standardizations* (ISO) rigorous MS review requirements to maintain certifications and assure continued SMS effectiveness. The documentation outline focuses on:

- The year's accomplishments in achieving sustainability policies, goals, objectives, and targets
- Identification of specific inabilities to achieve policies, goals, objectives, and targets with the least cost, effort, and risk
- A summary of lessons learned from the past year's portfolio of capability creation and performance improvement projects and initiatives
- Status of actions taken in response to previous SMS annual reviews
- A summary of findings obtained from internal and external conformance, conformity, and compliance audits, inspections, assessments, and stakeholder surveys conducted since the last annual SMS review
- Changes in the organization's operating environment that affect the SMS, including
  - New or increasing internal and external risks and opportunities, including regulatory changes
  - Internal financial, administrative, and operational changes, and
  - Other change-related opportunities for improving the SMS.
- Requirements for new SMS needs assessments – e.g. materiality and other assessments, stakeholder surveys, formal audits, and strength-weakness-opportunity-threat analyses
- Adequacy of current SMS resources and anticipated needs for new ones, and
- A summary of the overall SMS's *opportunities for improvement* (OFIs) and associated accountabilities required to assure its continued ability to add value to the organization.

The SMS champion provides a draft summary report of annual SMS assessment findings and recommended improvement prescriptions to the CEO for review, amendment, and approval. After review, amendment, and approval, the CEO delivers it to the board of directors for oversight deliberations, policy enhancement, and guidance. Summary reports typically include the following:

- Critical evaluation of prior SMS administration improvement actions
- New SMS administration OFIs

- Recommendations for sustainability policy amendments
- Summary of new resources needed to support and improve SMS administration
- Recommended changes to the SMS's organizational structures and standard operating procedures, along with other improvements, and
- New SMS administration accountabilities.

The SMS champion on behalf of the SMS oversight group distributes the CEO's approved summary report on a need-to-know basis to internal and external stakeholders. Further, the SMS champion on behalf of the SMS group manages the implementation of SMS administration improvement prescriptions approved by the CEO along with any policy-level amendments required by the board of directors. These prescriptions include SMS improvement accountabilities levied on units, functions, departments, and individuals in the organization.

The annual overall SMS management assessment is the final *check* activity that triggers the final *act* activity of the SMS's current PDCA cycle. Once this critical assessment of the overall SMS is complete, the next annual PDCA cycle starts.

11.2 Internal and External Assessments, Inspections, Audits, and Surveys

Whether or not an organization has an SMS, most will already have experience with *check* activities, such as assessments, inspections, audits, and surveys. The ones used in an SMS are important determinants of a sustainability program's structural integrity and operational efficacy. The following sections describe external and internal needs assessments that broaden and deepen an SMS's administration *check* function. While some are scheduled, others are often unannounced or conducted as needed.

11.2.1 External Assessments, Inspections, and Audits

External *check* activities include those contracted by organizations, such as materiality assessments, stakeholder surveys, ISO certification audits, and other conformance, conformity, and compliance audits. They also include actions undertaken directly by external stakeholders, such as regulatory agencies and downstream customers that actively manage their value chain relationships. External *check* activities may include:

- Environmental, health, and safety conformance and compliance inspections and audits by federal, state, regional, and local regulatory agencies
- Financial audits by federal and state regulators
- Financial- and sustainability-focused materiality assessments

- International and industry-standard conformance, conformity, and compliance audits and assessments, including ISO certification audits
- Audits and assessments of upstream suppliers and downstream supply chain partners to confirm compliance with contracts and sustainability-related regulations and standards
- Equity holder and other stakeholder audits, assessments, and surveys addressing sustainability and ESG-related transparency issues, including criticisms, and
- Investigations of and responses to public concerns and controversies.

11.2.2   Internal Inspections, Audits, and Assessments

Most sizeable organizations have nonfinancial internal audit and assessment functions to determine conformance, conformity, and compliance with regulatory and industry standards. They also conduct employee attitude and satisfaction surveys. They recognize that despite the potential for *audit weariness* due to their number and frequency, audits, assessments, and surveys are fundamental risk management activities. They enhance conformance, conformity, and compliance performance to protect the organization when properly executed.

Internal conformance, conformity, and compliance audits and assessments are often conducted before any externally imposed reviews. Such internal *check* activities enable an organization to discover and correct deficiencies to avoid sanctions that external entities might impose. This is in keeping with the three principal leadership responsibilities, i.e.:

- Make the numbers
- Run the organization without surprises, and
- When either of the first two isn't working, improve performance.

Contracted internal audits provide the means to discover any possible surprises. Then, the root causes of those possible surprises can be minimized or eradicated, thereby improving performance before any externally imposed inspection, audit, or assessment. As a result, these and other kinds of risks can be avoided:

- Regulatory and legal sanctions
- Contract breaches
- Operations disruptions leading to cost and expense increases
- Revenue reductions
- Diminished competitive position, and
- Damages to reputation.

Organizations that need to create or improve an internal nonfinancial audit and assessment function must recognize that it is imperative in a maturing enterprise. Although the inspection, audit, assessment, and survey function's design, implementation, and administration are beyond this book's scope, there are many publications on the subject. Two that are relevant to continuous improvement (CI) management systems are:

- *Auditing for Environmental Quality Leadership,* John T. Willig (Ed)
- *The Quality Audit Handbook*, ASQ Press Audit Division, J. P. Russell (Ed).

11.3 The Critical Importance of the SMS Check Function in Supply Chain Management

Check activities are increasingly important in assessing an SMS's soundness and efficacy. This is especially true regarding an organization's ability to monitor its external supply chain partners in the areas of contractual, regulatory, and industry standards conformance, conformity, and compliance. If warranted, a prospective or current supply chain participant with dependability, conformance/conformity/compliance, ESG, or public image issues is best avoided or removed from the supply chain when discovered.

The importance of identifying and removing poor performers or corrupt, unprincipled transgressors from the supply chain cannot be overstated. An effective inspection, audit, and assessment program will reveal problems or confirm that a supply chain participant is a trustworthy partner. Compared to the considerable expenses incurred during after-the-fact management of disruptive situations created by miscreant supply chain participants, inspections, audits, and assessments are essential low-cost risk management activities. As such, they are critical elements of the *check* and *act* functions of the PDCA cycle.

11.4 Chapter Takeaways

- The quarterly operations-level performance reviews of SMS capability creation and performance improvement projects detailed in Chapter 10 are only one aspect of an SMS's PDCA cycle *check* and *act* activities. The other is the annual high-level administrative SMS assessment. The annual SMS assessment is an important source of sustainability program performance information used in voluntary and mandated ESG reports.
- Given the functional differences between quarterly SMS reviews and annual assessments, the annual assessment should be conducted separately from the last quarterly review of the year.
- Critical to the success of any SMS is its annual administrative assessment, it is:
  - Conducted by the SMS oversight group or other chief-officer body

- Typically chaired by the SMS champion, and
- Attended by key organization unit, function, and department leaders and staff.

• Internal and external inspections, audits, surveys, and assessments are critical *check* activities that identify legal compliance and industry-standard conformance and conformity issues to avoid and reduce organizational risks and existing problems.

• The annual administrative assessment's detailed findings, OFIs, decisions, and SMS-improvement accountabilities are recorded, summarized, and disseminated to internal and key external stakeholders for follow up action.

• The annual administrative assessment is the end of the current year's PDCA cycle and the start of the next year's.

11.5 Further Reading

The first case study describes how a well-performing company conducts its quarterly and annual reviews. The second one demonstrates the importance of comprehensive assessments, audits, and inspections throughout a company's value system.

11.5.1 The Teck Resources Limited Red Dog Mine Management System Review Processes

Canadian mining company Teck Resources Limited (Teck) owns and operates the Red Dog Mine (RDM), the world's largest lead-zinc mine. Situated above the Arctic Circle in Alaska, it produces 10% of the world's zinc and over half of Alaska's annual mineral revenues.

Operating in North and South America, Teck has been globally recognized for its sustainability efforts. It has been included in the *Standard and Poor's Global Corporate Sustainability Assessment* for 14 consecutive years and was ranked fourth within the metals and mining industry sector in 2023. Further, the company was included in the *2023 Global 100 Most Sustainable Corporations* list by Corporate Knights for the sixth year.

Integral to the sustainability program is RDM's *ISO 14001 Environmental Management System* (EMS) *Certification*, first awarded and continuously maintained since 2004. Maintenance of the certification requires auditing and recertification by external auditors every three years, along with interim annual verification audits. Also contributing to Teck's sustainability program is RDM's implementation of the Mining Association of Canada's *Toward Sustainable Mining Safety and Health Protocol*.

Supporting its ISO 14001 certification, RDM conducts internal EMS efficacy and performance improvement reviews. Quarterly reviews of its CI environmental projects and annual administrative reviews of its EMS are conducted to assess and improve the system to better meet its own performance improvement needs and corporate requirements. All actionable tasks from the quarterly project reviews and annual administrative reviews are entered into the RDM's monitoring, tracking, and reporting system to formally establish accountabilities and exploit improvement opportunities.

Regarding the EMS's annual administrative review, RDM's Environmental Coordinator typically spends a week each year preparing for, conducting, and reporting the results. Additionally, the Environmental Manager spends a week reviewing the EMS's current slate of CI environmental projects. With preparations complete, RDM's general manager, senior management team, operations-level managers, and select supervisors attend the half-day review meeting. Operations managers prepare and deliver presentations regarding their departments' environmental performances for the previous year. Representative concerns and agenda items include:

- The year's accomplishments
- Identification of specific inabilities to achieve policies, strategic goals, and tactical objectives
- Lessons learned from the past year's performance improvement projects and initiatives
- Status of actions taken in response to previous EMS annual reviews
- A summary of internal and external compliance audits, inspections, assessments, and stakeholder surveys conducted since the last annual EMS review
- Changes in the organization's operating environment that affect the EMS
- Requirements for new EMS needs assessments, e.g. SWOT analyses and formal audits
- Adequacy of current EMS resources and anticipated needs for new ones, and
- A summary of overall EMS opportunities for improvement, associated accountabilities, and continued ability to add value to the organization.

The annual review meeting is recorded on an *Annual Management Review Record Form*, a summary report archived in the RDM's documentation tracking system for site-wide availability. It is also distributed to the top management team and other attendees.

In summary, the following activities are critical features of RDM's EMS that contribute to Teck's sustainability program success:

- Top management and all departments actively participate in the reviews
- Environmental improvement projects are closely tracked through quarterly reviews

- The review findings and resulting accountabilities are formally documented and widely distributed internally and externally, plus they are included in internal training programs
- Review accountabilities are recorded, tracked, monitored, and reported
- The quarterly and annual reviews ensure that operations-level environmental performance and EMS administration are continuously improved, and
- External annual and tri-annual audits to maintain their ISO 140001 certification provide frequent objective assessments of EMS efficacy.

Sources:

- **Teck 2023 Sustainability Report, www.teck.com/media/2023-Sustainability-Report.pdf, Retrieved 2 July 2024**
- **Questionnaire Provided by the Authors to and Completed by S. Staley, et al, Teck Resources, 8 July 2024**
- **Grosskopf/Orion ISO 14001 RDM verification audits 2019, 2020, and 2021**

11.5.2 Perdue Farms Child-Labor Case: Subcontractor's Alleged Violations of Child-Labor Laws

This case study illustrates the importance of sustainability program assessments, especially regarding internal audits, inspections, assessments, and regulatory actions. It describes what happened when one industry leader, Perdue Farms, unsuccessfully managed its supply chain in conformance with its established sustainability standards and compliance with external legal requirements.

According to a *New York Times* article on 18 September 2023, the United States Department of Labor alleged that one of Perdue's subcontractors, Fayette Janitorial Service LLC, violated child-labor laws by hiring middle and high school-aged children to clean meat-processing equipment during overnight shifts at a plant in Accomac, Virginia. This work, during which at least one child was seriously injured, involved handling acids and pressure hoses to remove blood and tissue scraps from industrial-scale machines. Further complicating supply chain management issues is the plant in question is not directly owned by Perdue, according to a 6 December 2023 article posted by *JURIST News*. Instead, it belongs to a smaller independent contractor.

This and other enforcement actions are part of efforts initiated in February 2023 by a Federal *Interagency Task Force to Combat Child-Labor Exploitation* to protect children from exploitative situations in the workplace. This appears to be the first time the Department of Labor has attempted to hold a prime contractor jointly liable for the child-labor violations committed by a subcontractor. The legal basis is the *joint employment doctrine*, a legal principle that when one employer exerts enough control over the working conditions of its subcontractor's employees, it may be considered legally liable for those workers.

Perdue Farms is a well-respected century-old family-owned business that declared in its *2023 Company Stewardship Report*: "… we firmly believe that responsible business practices, a diverse workforce, and ethical treatment of our animals are not only the right things to do but also critical for long-term success." Further, the report stated that the company's trademarked vision is: "To be the most trusted name in food and agricultural products."

In response to the allegations, a Perdue spokesperson contacted during the Department of Labor's investigation stated:

- The company plans "to cooperate fully with any government inquiry on this matter"
- The company has "strict, longstanding policies in place for Perdue associates to prevent minors from working hazardous jobs in violation of the law"
- "We hold our sanitation contractors to the same high standards"
- "We are conducting a comprehensive third-party audit of child-labor prevention and protection procedures, including a compliance audit of contractors" and
- The company will take "appropriate actions" based on the findings.

However, the problem and its lesson are that after-the-fact actions cannot repair the damages that preemptive ones prevent.

A typical supply chain risk management tactic is periodically auditing suppliers and subcontractors, especially regarding social and environmental concerns. Two exemplary comprehensive supplier and subcontractor audit programs are Target's *Responsible Sourcing and Sustainability Audit Program* and Nike's audits of its *Standards of Compliance*. Although it has limited-topic audit programs, at the time this book was written, Perdue Farms did not have a comprehensive supply chain audit program like these.

A critical annual management review of Perdue's sustainability program, including benchmarking other companies, could have identified the need for a more stringent supplier and subcontractor audit program. However, that is now a moot point. The Department of Labor's investigation and enforcement actions have resulted in unnecessary legal defense expenses and may lead to fines and other sanctions in the near term, along with long-term reputation damage.

This case study highlights essential sustainability preventative principles and practices described in this book, including the following.

- Social responsibility, environmental stewardship, and other supply chain risks have significant and costly disruption potentials.
- Therefore, supply chain risks and their remedies must be a primary concern during annual management assessments of sustainability programs.
- The importance of benchmarking other companies' sustainability programs to proactively define opportunities for sustainability program improvement cannot be overstated.

- Adverse supply chain issues can occur anywhere in a supply chain's upstream, midstream, and downstream parts. Pay attention to each.
- Companies must set, communicate, and continually emphasize contractually binding social responsibility and environmental stewardship performance standards for all supply chain participants.
- To ensure that the participants continue to operate within those standards, periodic internal and external supply chain audits and other due diligence efforts must be rigorous and proactive.
- The discovery of adverse supply chain issues requires decisive legal and ethical actions to protect company interests, including severing miscreant supplier and service provider contractual relationships.
- Problems in a company's supply chain can perceptually render heretofore legitimate claims of performance quality, environmental stewardship, and social responsibility to be nothing more than hypocritical greenwashing.

Sources:

- https://www.nytimes.com/2023/09/23/us/tyson-perdue-child-labor.html
- https://news.bloomberglaw.com/daily-labor-report/perdue-tyson-foods-face-unique-probe-in-child-labor-crackdownwww.dol.gov/agencies/whd/data/child-labor
- www.jurist.org/commentary/2023/12/child-labor-investigation-at-tyson-foods-inc-is-supply-chain-due-diligence-the-next-step/
- www.hklaw.com/en/news/intheheadlines/2023/10/perdue-tyson-foods-face-unique-probe-in-child-labor-crackdown

Retrieved: 29 May 2024

Appendix A

Origins and Evolution of the SMS Model

This appendix describes the origins and evolution of the book's *continuous improvement* (CI) sustainability management system (SMS) model.

A.1 Introduction: Where We Have Been and Where We Are

The SMS model's origins are not rooted exclusively in sustainability. More broadly, this model is a sustainability-focused application of advancements in CI management concepts that have evolved over decades.

Although initiated in the 1920s with a statistical process control perspective, widespread adoption of the CI methods developed and promoted by luminaries such as Walter Shewhart, Joseph M. Juran, and W. Edwards Deming was limited before 1951. However, in 1951, the Japanese Union of Scientists and Engineers (JUSE) awarded the first *Deming Prize*, launching that country's quality revolution that eventually spread throughout the industrialized nations.

Decades later, 1987 was a milestone in further adopting quality management concepts. It was also the year sustainability was first globally recognized as a paramount principle in managing activities affecting people and their environments. Here are the events that made 1987 notable:

- The International Organization for Standardization (ISO) published its first management system standard, *the ISO 9001 Quality Management System* (MS) Standard;
- The *Malcolm Baldrige National Quality Award Program* was launched in the United States to codify performance standards for organizational quality; and
- The United Nations capped off the year's achievements with its *Brundtland Report*, which first defined sustainability and incubated triple-bottom-line organizational management concepts.

Sustainability Programs: A Design Guide to Achieving Financial, Social, and Environmental Performance, First Edition. William Borges and John Grosskopf.

However, none of these early events provided specific management processes to comprehensively improve financial performance, reduce environmental impacts, and provide social benefits. Early attempts often failed to achieve single-issue goals, let alone comprehensive organizational sustainability.

In the 2020s, it is apparent that comprehensive and adaptable models for operationalizing sustainability are urgently needed to improve sustainability's triple bottom lines with the least cost, effort, and risk. With its emphasis on organizational development (OD), this book's model is designed to do just that. It focuses on developing and achieving sustainability policies, strategies, tactics and targets within a systematic and systemic CI MS employing proven concepts and methods. Its OD perspective provides a stepwise process to create and manage a customized SMS in any organization with the need, will, perseverance, and resources. It is not a perfect model by any means. However, organizations that use its concepts and methods can create the next generation of noteworthy SMS refinements.

A.2   In the Beginning: The 1960s and 1970s

The direct developmental path to this book's SMS model stretches over six decades. It started with the establishment of NASA's R&QA Office in 1962. The R&QA Office began specifying and managing performance requirements through NASA contracts by consolidating programs from earlier military and civilian space quality assurance programs. A point of pride for co-author John Grosskopf, the first contractor involved was his early employer, *General Dynamics* (GD). The effect of such contracts throughout the aerospace industry was profound as companies began adopting the systematic management methodologies promoted by the *total quality management* (TQM) pioneers noted above. In the case of this SMS model, many of the core processes were initially developed by another NASA contractor, TRW Aerospace.

A.3   Quality Assurance Morphs into CI Organizational Management Systems – The 1980s

Later in their careers, former TRW Aerospace executives were involved in investment banking at *Hambrecht and Quist* (H&Q) in the San Francisco Bay Area. The firm focused on the technology and Internet sectors in Silicon Valley. In addition to underwriting initial public offerings (IPOs) for Apple Computer, Genentech, and Adobe Systems in the 1980s, H&Q also backed the IPO of Amazon.com Inc. In response to start-up companies' stark business realities, the firm also provided financial turnaround and organizational restructuring services to protect investor interests. After securing financial control of troubled companies, H&Q imposed rigid, highly disciplined planning and control processes to regain profitability within two fiscal quarters. Once achieved, that profitability would be sustained over the next year. Those planning and control processes were derived from the H&Q executives' experience at TRW Aerospace.

One of the notable companies H&Q restructured in the 1980s was the medical imaging pioneer ADAC Laboratories. Thanks to its advanced MS, ADAC won the *Malcolm Baldrige National Quality Award* the first year it applied in 1996. Mr. Borges participated in the introduction and development of ADAC's new MS during its financial turnaround by restructuring sales and operations departments. Due to its highly effective MS, financial performance, and advanced products, ADAC was later acquired and absorbed into medical device giant Philips Medical Systems.

In the late 1980s, GD focused management efforts on improving production efficiencies to eliminate waste in all its forms with an initial emphasis on high-risk, high-cost hazardous wastes. Based on CI MS principles and practices, these efforts became known throughout the aerospace industry as GD's *Zero-Discharge Program*. Its most effective implementation was at GD's Valley Systems Division, the producer of the Stinger missile system, where Mr. Grosskopf was the Environmental Program Manager. With the support of the Division's executive team, W. Edwards Deming was retained to transform the quality program – including its environmental program elements – by introducing TQM methods to create an early EMS. Through that EMs, the Division achieved practicable zero-waste and zero-adverse emissions goals within three years through this transformation. Also, during this time, it designed and built a closed-loop production water system to eliminate aqueous discharges. Concomitantly, Stinger missile production increased four-fold to meet customer requirements.

GD proactively promoted its TQM-based environmental programs by sharing them throughout the aerospace industry, especially with those companies operating in California, through the *California Aerospace Environmental Association* (CAEA), where Mr. Grosskopf served as chair for two years. The company was also instrumental in launching the *San Diego Industrial Environmental Association*.

A.4   CI Management Systems Become Green: The 1990s

In the 1990s, interest in CI MSs grew, and applications became more widespread, sophisticated, and effective. The *International Organization for Standardization's* (ISO) work codifying CI concepts – especially its pioneering *ISO 90001 Quality Management System Standard* in 1987 – significantly accelerated awareness and acceptance of these new ways to manage organizations. This new awareness and acceptance was so widespread that even the United States Navy was emboldened to develop and roll out its version of CI principles and practices, the *Total Quality Leadership (TQL)* Program.

In the earliest days of the Navy's initiative, United States Marine Corps Base Camp Pendleton found itself the focus of an *Environmental Compliance Evaluation* (ECE) mandated by the United States Marine Corps Headquarters. The ECE findings did not complement the Base's considerable efforts to manage its environmental responsibilities. In response, the leaders and staff were eager to apply their new elementary knowledge of high-level TQL concepts to resolve the ECE issues. However, they had a major stumbling block; they did not know how to do it. Through his employer, Mr. Borges was assigned to help the Base's leaders find a solution. Using his in-depth knowledge of organizational turnarounds and CI MSs gained at ADAC Laboratories, he designed a TQL-formatted EMS that enabled:

- Prioritization of ECE issues
- Design of issue-resolution projects
- Development of program budgets
- Acquisition of personnel, funding, and material resources, and
- Definition of project tracking, control, and evaluation processes.

The Base leaders successfully executed the EMS without significant complications. By the end of 1992, Camp Pendleton had set the benchmark for the Marine Corps and Navy, plus the entire United States Department of Defense (DoD). Camp Pendleton's proof-of-concept success contributed to the current requirement for EMSs at all DoD facilities.

A.5   Time Out for Reflection in the 1990s

The next step in developing this book's SMS model occurred in the late 1990s in the form of Mr. Borges' master of business administration thesis, *A Model EMS Development Plan*. The Camp Pendleton EMS had been a relatively simple matter of coaching the base leaders and staff to create an EMS in a tightly run command-and-control government organization. However, the thesis described a different kind of effort. It detailed a change management process to design and introduce an EMS in a for-profit corporation. The following concepts emerged from the thesis effort:

- All CI MSs share the same basic structures, processes, and disciplines irrespective of the topics used to develop overarching policies, broad strategies, specific tactics and targets, and performance improvement initiatives.

- Technical professionals – e.g. engineers, scientists, technologists, clinicians, and IT specialists – typically know little of the intricacies of organizational management, compounded by the problem that management professionals know little of technical specialties.
- The design and implementation of a specialty CI MS in a conventionally run *management-by-objectives* (MBO) organization requires extraordinary change management efforts. This is because CI skill sets, such as mastery of contemporary process-improvement methods, are often underdeveloped in MBO organizations.

A.6 The 21st Century Part 1: CI Management Concepts Applied in a Tough Environment

After academic refinement of the model's OD processes, it was time to apply and test them. The testbed was a regional integrated healthcare system operating under an inadequate MBO MS overdue for systemic change. The healthcare system's financial performance was barely profitable, and its clinical performance was middling. Mr. Borges and a colleague from ADAC Laboratories were hired to address administration and clinical performance issues. Initially, they assessed the MBO MS shortcomings while coaching department leaders on the rudiments of process improvement, i.e. finding root causes of often life-threatening problems and designing and completing corrective actions that produced measurable results. They then helped introduce the foundational processes of a CI MS. Specifically, these processes included:

- Whole-house instruction for executives and department leaders in contemporary process-improvement methods
- Quarterly systematic definition and prioritization of the organization's most pressing needs
- Creation and introduction of an automated system to record, track, and evaluate progress and success of process-improvement accountabilities, and
- Introduction of quarterly whole-company reporting and evaluation events focused on unit, function, and department performance.

Although the healthcare system did not fully adopt the CI MS model, it significantly modified its old one into a hybrid combining MBO and CI elements. The MS changes received recognition as industry benchmarks from *The Advisory Board Company*, one of healthcare's leading best practices research-and-education organizations. Other performance improvement results included:

- A 15-fold increase in annual profits over four years

- A 27% increase in patient satisfaction scores, and
- A *"Top 100 Healthcare Network"* rating six years in a row by the healthcare rating organization *Verispan*.

A.7 The 21st Century Part 2: Getting Greener

Mr. Borges wrote a blog, *Creating Hospital Sustainability Programs*, in the early 2000-teens. The blog described a step-by-step OD design and implementation process for healthcare professionals considering implementing sustainability programs. The blog attracted nearly 25000 readers by focusing on the difficulties aspiring sustainability change agents face in healthcare. In addition to providing a program design process, it offered advice for structured change management and methods for dealing with obstructionists.

During the late 1990s and early 2000s, Mr. Grosskopf launched a highly successful consultancy, *The ISONetwork*. The firm was among the earliest focusing on designing and introducing topical CI-based MSs, including quality, environmental, health and safety, and even security. The firm developed essential concepts and methods to combine topical MSs into some of the earliest multiple-topic integrated SMSs. The ISONetwork introduced hundreds of clients in the public and private sectors to CI-based MSs and assisted many in securing a variety of ISO certifications.

Notably, Mr. Grosskopf led and participated in developing several pioneering CI-based MSs. These included the *United States Environmental Protection Agency's National Enforcement Investigations Center*, which retained him to create the compliance-focused EMS model still used in alternative enforcement cases, such as *Supplemental Environmental Projects* (SEP) required by consent decrees and actions. Following the EMS model's development, he was retained to assist with its first application in a landmark enforcement action against a major mining company that was enjoined by several state attorneys general.

A.8 The 21st Century Part 3: Those Who Can Also Teach

The co-authors' efforts in teaching aspiring students and professionals have helped transform organizations into more productive, profitable, and sustainable ones. As adjunct professors, guest lecturers, and corporate trainers, Messrs. Borges and Grosskopf taught management and sustainability courses at the undergrad and graduate levels for colleges and universities, plus specialty courses for corporations and industry associations. Based on the instructors' academic preparation and first-hand experience, these courses stressed systematic change management methods.

In the late 2000s, Mr. Borges designed two undergraduate courses, *Environmental Management Systems* and *The Sustainable Organization,* for a university system serving working adults. During this time, Mr. Grosskopf developed and taught a CI-based environmental, health, and safety (EH&S) MS course at the *Rocky Mountain Education Center* at *Red Rocks Community College's OSHA Institute for Education.* This was one of the first – if not the first – accredited ISO-principle courses in the United States.

In the mid-2000-teens, the *California Community Colleges Chancellor's Office* awarded a sizable grant to the *Grossmont-Cuyamaca Community College District* to develop and deliver a professional certificate program in *Sustainable Supply Chain Management* for mid-career technical and management professionals. Mr. Borges was retained to design and teach the curriculum. The courses were:

- Sustainable supply chain management
- Change management
- Project management, including preparation for the *Project Management Institute's Certified Associate in Project Management* exam, and
- *LEED Green Associate* exam preparation.

Curriculum development and delivery provided further opportunities to refine the SMS model, especially the expansion and refinement of descriptive materials originally appearing in the earlier hospital sustainability blog.

A.9 The Evolved Sustainability Management System Model

Starting in the 1920s, evolving MS CI processes have become increasingly effective in systematically achieving organizational policies, strategies, tactics, and targets with the least cost, effort, and risk. These processes provided a starting point for the co-authors to consider the desirability and debate the shortcomings of various principles and practices that should be included in an ideal SMS model. The resulting book reflects the best design and management advice the co-authors can provide any reader who aspires to be their organization's sustainability champion or a key participant in a sustainability program.

Appendix B

Essential Change Management Concepts

B.1 Overview of Change Management

The authors have learned the importance of change management (CM) by leading and participating in scores of management system creation and improvement initiatives. Their key lesson is that ignoring formal CM is one of the surest ways to underperform or fail with any initiative, especially anything as complex as a *sustainability management system* (SMS).

At its simplest:

> *Change management helps organizations, groups, and individuals create new organizational capabilities or improve existing ones.*

A more complete definition of CM is:

> *Planning, initiating, realizing, controlling, stabilizing, and sustaining new and altered work activities at the organizational, group, and personal levels.*

CM involves a balance of considerations from organizational management, industrial engineering, and psychology disciplines. It has two distinct yet equally important perspectives: the perspectives of *top-down managers* and *bottom-up employees*. Because front-line employees execute day-to-day activities, they make the organization's systems and processes come to life. Therefore, their unique, practical perspectives are essential to the change initiative's success.

Further, there are two CM focal points: *organizational change* and *individual change*. The former focuses on broad practices and skills to help the organization understand, accept, and support the needed change. The latter focuses on these ideas:

- Organizations do not change until individuals do

Sustainability Programs: A Design Guide to Achieving Financial, Social, and Environmental Performance, First Edition. William Borges and John Grosskopf.

- No matter how large the initiative, its success depends on each employee changing the way they work, and
- Effective CM requires a focus on how each person successfully makes a change.

Along with this focus, four CM principles come into play during a change initiative:

- Everyone reacts differently
- Everyone has intellectual, emotional, and material support needs
- Emotional reactions – including perceptions of loss and their attendant fears – must be expected and managed, and
- Positive and negative outcome expectations must also be managed.

The high-level question that initiates and sustains genuine change is:

> *Why is change needed?*

Its full answer requires answering these follow-on questions:

- *What is the point of what we are currently doing?*
- *What has to "die" before we can move to something new?*
- *What are the fundamental values of our products and services? And in whose eyes?*
- *What personal meaning do people find in what we are doing?*
- *What would happen if we did nothing?*
- *What are the capacities and strengths that we are not using entirely?*
- *And, regarding sustainability, what are we leaving for the next generation, i.e. what will our legacy be?*

Once the need for change and its implications are clearly understood, common goals for CM activities include:

- Learn and capitalize on the benefits of change
- Identify approaches to change that create opportunities, not problems or crises
- Acknowledge and address previous difficulties and losses
- Broaden awareness of the change process
- Avoid victimization
- Convert tensions and fears into confidence and excitement, and
- Establish positive, adaptable internal dialogues.

B.2   Change Management Success and Failure Factors

This point has been made throughout the book:

> *The creation of an SMS is a major strategic initiative.*

Further, because sustainability is still a relatively new, unfamiliar, and even exotic idea for many in the ranks and files, creating an SMS is a radical organizational development. As such, success factors for radical transformations include:

- A clear shared vision
- Strong, determined leadership to see it through
- Determined and continuous consensus-based catalytic activity at the chief executive (CxO) level
- A comprehensive and systematic approach
- Proactive and trustworthy internal communications, both top-down and bottom-up
- Interactive communications with affected external stakeholders
- Adequate budgetary and other resources
- Extensive education, competency-based instruction, and task-specific training at all levels
- Permanently empowered employees, and
- Most employees accept the change.

Working against success are such change failure factors as:

- No compelling vision
- No clear, unwavering, high-level mandate
- No powerful guiding coalition
- No formal accountabilities
- No meaningful incentives
- No sense of urgency
- Inadequate resources
- Failure to adequately communicate the vision with internal and external stakeholders
- Failure to anchor changes in organizational structures to change the culture
- Failure to empower others to act
- Obstacles permitted to block the vision
- Complacency is tolerated
- Failure to build on accomplishments
- Failure to create and celebrate short-term wins
- Failure to institutionalize and thereby sustain the change, and
- Victory is declared too soon.

B.3   Schools of Thought within Change Management

There are two interrelated CM schools of thought: the *engineering school* and the *psychology school*. The engineering school focuses on observable, measurable organizational structures that can be changed or improved. Unfortunately, organizations that primarily embrace the engineering approach often do not consider the social aspects of CM until their projects encounter resistance or other serious personnel problems during implementation. When this occurs, many organizations' approaches to the social aspects of CM is *ad hoc*. Compounding the problem, engineering perspectives tend to isolate the *people* problems from the *process* ones and then try to eliminate the former with a quick and often ineffectual fix.

In contrast, the psychology school is concerned with how humans react to their environment. It focuses on how an individual thinks and behaves when confronted with change. This school's key concepts include:

- CM is primarily a relationship activity
- No matter how research-based or technical a project, it will always reach a point at which the success of work will hinge on the quality of the relationship the change agents have with the stakeholders
- Change implementation cannot be treated as a fundamentally rational process even though tasks can be structured
- People choose to commit to change based on emotion, intuition, trust, faith, hope, and personal values, and
- Any change requires a shift in what is tangible – such as structures and methods – and intangible – such as relationships and commitment.

The extreme application of either of these two schools of thought in isolation will be unsuccessful. An exclusively engineering approach will result in effective solutions that risk inadequate implementation. Conversely, an exclusive psychology approach will help make groups and individuals receptive to new things without an appreciation, understanding, or structured means to implement and institutionalize a successful change. Therefore, the engineering and psychological approaches must be integrated into a comprehensive change process.

B.4   Essential Change Management Concepts

Despite numerous competing CM approaches, there is a simple underlying concept, as shown in Kurt Lewin's model, *Three Phases of Change Management,* Figure B.1. Lewin's model includes these three common sets of emotion-based change-response phases to new initiatives:

- The *SARAH* process: *shock, anger, rejection, acceptance,* and *healing*
- The *Kübler-Ross Five Stages of Grief Model*: *denial, anger, bargaining, depression* and *acceptance*, and
- In team building, *forming, storming, norming,* and *performing.*

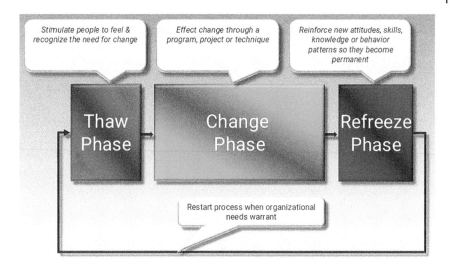

Stimulate people to feel & recognize the need for change

Effect change through a program, project or technique

Reinforce new attitudes, skills, knowledge or behavior patterns so they become permanent

Thaw Phase

Change Phase

Refreeze Phase

Restart process when organizational needs warrant

Figure B.1   Kurt Lewin's. Three Phases of Change Management.
Source: W. Borges.

These emotion-based processes are mentioned here because they summarize the reactions and behaviors that must be managed during the structured *thaw, change,* and *refreeze* phases of the Lewin model. Never trivialize these very real and consequential responses to change. Instead, prepare people to anticipate and understand what they will be experiencing and support them while they work through their individual and collective feelings.

B.5   The Role of a Change Agent

A change agent promotes and enables change at the individual, group, and organization levels. There may be a single or multiple change agents for any given initiative. It all depends on the change's size, nature, and complexity. At their best, change agents must go beyond the tasks associated with the change and engage in mutual learning to exchange ideas with individuals and groups. These social interactions are intended to lead to insights, resolution of differences, and ultimately, accomplishment of the desired change.

It should be no surprise that change agents have multiple roles. However, change agents must avoid such power-diminishing labels as *friend, crony, underling,* or *sycophant.* Here are the main roles:

- Expert
- Diplomat

- Advocate
- Instigator
- Investigator
- Facilitator
- Educator
- Counselor
- Mediator, and
- Enforcer.

Change agents must exercise various organizational authorities and influence to fulfill these roles. In Chapter 6, organization leaders are encouraged to create *chief sustainability officer, sustainability vice president,* or *corporate sustainability director* positions to serve as the SMS champion, the lead change agent in a sustainability program. Leaders in such high-level positions should be able to exercise most, if not all, of the following authorities:

- *Legitimate,* aka *Position,* A*uthority*: The responsibilities defined for an organizational role
- *Referent Authority*: In the case of change management, it is an ability of a change agent to influence people because of their admiration and respect for or identification with someone who supports the change agent
- *Expert Authority*: A high level of knowledge or a skill level that others in an organization see, need, and want
- *Informational Authority*: An ability to access information to influence or evoke certain behaviors within others
- *Coercive Authority*: The authority to detect and sanction undesired behaviors, and
- *Reward Authority*: The ability to reward others.

There is a widespread misconception that with their authority, a change agent is solely responsible for the change. It is simply not true and misguided to think so. The ultimate success of a change initiative resides with all of the organization's leaders and personnel. Despite this reality, change agents are too often expected to be *messiahs* on whom the whole responsibility for initiative success is placed. Too frequently, those who mistakenly assume a change agent will perform miracles sit back and watch from the sidelines. In their minds, is that not what an expert change agent is supposed to do? This type of thinking is a significant contributor to underperformance and failure.

B.6   Change Agent Dos and Don'ts

This brings the discussion to change agent *dos* and *don'ts*. Here are some *dos*.

- Remember at all times that you are a trusted partner in the change effort who has personal, ethical, professional, and legal responsibilities to the organization.
- Do reinforce the idea that problem solutions are essential to the organization.
- Do emphasize the need for urgency.
- Look for, find, and use hidden champions.
- Once recognized, do start slaughtering *sacred cows* – i.e. those attitudes, issues, or systems that are never objectively challenged, critiqued, or corrected – with diplomatic assertiveness.
- Do adequately prepare for change. Problem solutions come later in the process.
- Do keep asking: *What will you do differently next time?*
- Do remember that *people don't know what they don't know.* For significant changes like SMSs, that can be quite a lot.
- Do remember that as the change initiative progresses, a change agent's technical knowledge remains indispensable. However, their transformational leadership skills rise in importance.
- Do keep asking yourself and those affected by the change: *Are you getting what you want?*
- Do support other leaders in their efforts to complete the initiative successfully.
- Do confirm positive expectations and manage negative ones.
- Do use language that is accurate, matter-of-fact, and non-punishing, especially when giving feedback.
- Be diplomatically assertive around facts, data, analysis, outcomes, alternatives, and recommendations based on objective pros-and-cons evaluations.
- Do pushback. However, that pushback must be respectful, fact-based, and data-driven.
- As strange as it may sound during a change initiative, do give people permission to change.

Fortunately, the *don'ts* list is not as long but equally important.

- Don't be aggressive, but do be diplomatically assertive.
- Certainly, don't be nonassertive. Even if obstructionists wear you down, avoid this attitude: *Anything you want to do is okay by me.*
- Don't include any *you dummy* implications in your responses to challengers.
- Don't take anything personally.
- Don't support stances that reduce the organization's problem-solving ability.
- Other than a passion for successful change, do not overly project your emotions.
- Don't provide knee-jerk solutions to ill-defined problems.
- Don't be judge, jury, prosecutor, defendant, witness, or finger-pointer when problems arise. Just help define the root cause and corrective action without punishing people for the sins of the work process.

B.7 Preparations for Change

To start the change buy-in process, all stakeholders must know how the proposed change satisfies the following criteria.

- *Congruency:* The change must align with the organization's values, vision, and mission.
- *Benefit:* The change should benefit the people asked to change. It should be perceived as *"a better way"* that answers the critical question: *What's in it for me?* (WIIFM).
- *Compatibility:* The change should be as compatible as possible with the existing values and experiences of the people asked to change.
- *Complexity:* The change should be no more complex than necessary. It must be as easy as possible for people to understand and use.
- *Try-ability:* Break complex activities into simple tasks. The change should be something that people can try step-by-step and make progressive adjustments in their activities.

B.7.1 A Basic Change Process

The list of activities below provides a framework within which CM concepts, including *dos* and *don'ts* are integrated. This book's SMS design, implementation, and management processes incorporate these activities. Further, note that these activities are embedded in the *Plan-Do-Check-Act* (PDCA) cycle. This process is helpful because it can be scaled up for any organization-wide initiative and focused down to discrete project levels.

- Assess the need for change.
- Set a policy-level mandate for change based on organizational needs.
- Recruit the CM team.
- Complete a change readiness assessment.
- Secure resource commitments to define, implement, and sustain the necessary changes.
- Develop the CM plan:
  - Define and design the tasks, and
  - Identify likely resistance and prescribe obviations and mitigations.
- Levy appropriate accountabilities on responsible executives, staff, units, functions, and departments.
- Formally, continually and intensely communicate:
  - Management intentions
  - Change progress, and
  - Immediate successes and long-term change-sustaining efforts.

- Provide ongoing effective education, competency-based instruction, and task-specific training to prepare for, accomplish, and ultimately institutionalize the change.
- Formally kick off and complete the change initiative.
- Take advantage of *simple tests of change* and other *quick-win* initiatives.
- Actively manage the change initiative programmatically and in the moment by monitoring, detecting, and correcting performance issues, including managing obstructionist behaviors.
- Frequently measure and report progress against high-level policy mandates, strategic goals, tactical objectives and targets, project plans, and other initiatives.
- Define and apply lessons learned to improve the CM process and institutionalize the changes.
- Lastly, formally celebrate the change effort and its successes, rewarding those who made it happen.

B.7.2 Communications Planning

Internal and external stakeholders must be well informed about the purpose of a change and provided opportunities to share their ideas about its implementation. To do otherwise risks failure at the outset. Further, communications must be tailored to the unique perspectives of each stakeholder group with a clear focus on the WIIFM concept. However, it must be recognized that over-communication is just as undesirable as under-communication.

Communications plans must, at a minimum, address change from the stakeholder groups' WIIFM perspectives in a *what, why, when, where, who,* and *how* plus *check* (5Ws & 1H + Check) format. Like other change plans, communications plans must be documented, subjected to periodic reviews, and improved. An example of a communications plan worksheet is shown in Figure B.2.

The matrix helps design individual communication plan elements for each of the various stakeholder groups:

- Board of directors members, CxOs, and other executives
- Mid-level managers and supervisors
- Employees
- Suppliers and contractors, and
- Customers.

Additionally, organizations should consider, as appropriate, the following stakeholders in communication planning efforts:

- Regulators
- Advocates
- Affected communities, and
- Other significant external stakeholders.

Target stakeholder group:					
Timing	Message content	Delivery mechanism	Sender	Frequency, dates, & times	Message risks & counter-measures
Opportunity-for-improvement discovery & "go" decision					
Before & after readiness-for-change assessment					
During the change design					
Before change implementation					
During change implementation					
After change implementation					

Figure B.2 Communications Planning Matrix.
Source: W. Borges

As noted above, there are risks in CM communications. Groups of people can exhibit these behaviors when stressed, which require obviation and mitigation measures:

- They are often overwhelmed by the new work process inputs, including information from the communications process
- Further, they are often confused by the requirements of the new input-transforming work processes, and
- As a result, they work hard, yet, for all their efforts, they are at risk of producing ineffective results.

A root cause of CM communications problems is the inability to acquire and analyze information to make clear-cut decisions under stress and uncertainty. Effective problem-prevention messaging tactics include the following:

- Assure timeliness in message delivery
- Remain consistent in the messages and among the messengers
- Apply the *KISS* principle, i.e. *keep it simple, stupid*
  - Present information in plain terms
  - Avoid complexity, and
  - Avoid ambiguity and mixed messages
- Scrutinize every message for innate *silliness* before distribution, and

- Monitor, detect, and immediately correct for confusion among target audiences and individuals.

Be aware that the readiness assessment processes should only be viewed as planning aids, not as the bulk of the CM effort. Two critical readiness assessments are needed at the onset of the change:

- Assessment of the change itself and
- Assessment of the organization, particularly its leadership.

Readiness assessments are used for:

- Defining risks to the change effort and identifying potential obstacles
- Determining if any unique tactics are necessary to support this change, and
- Customizing and amending change-implementation plans, communication plans, knowledge-transfer plans, and related activities.

The assessment of the change itself examines its scope, depth, complexity, and overall scale. Specific items that should be addressed during readiness assessments are:

- The types of employees and the number that will be impacted
- The scope of the change, including considerations of affected workgroups, departments, functions, units, and entire enterprises
- The expected duration of the change effort from the old way to the institutionalized new way
- The type of change, such as process, technology, organization, job roles, merger, and strategy
- The amount of change from where the organization is currently to where it needs to be, and
- Other internal and external change factors specific to the organization.

Once the nature of the change has been characterized, it is time to assess the organization's readiness for change. The assessment includes the following tasks.

- At the outset, determine top management's understanding and appreciation of the change, their predisposition and commitment, and recognition of the effort required for the whole organization's transformation.
- Identify the enabling and obstruction elements of the organization's tangible structures and intangible culture.
- Identify and evaluate the organization's capacity for change by cataloging all current and upcoming initiatives.

- Evaluate ways essential leadership skills, styles, and power distribution may enable or obstruct change. A critical consideration in evaluating leadership styles is assessing the amount of competency that resides within those styles. Some individuals emphasize the style of their activities to conceal their inadequacies. So, look for the *Dunning–Kruger Effect* in people with low ability, expertise, or experience who overestimate their abilities. At the same time, identify high performers who underestimate their talents.
- Determine middle management's predisposition toward the change and define methods to gain and maintain their support.
- Define residual beneficial and adverse effects of past changes, plus prescribe actions to leverage favorable outcomes while avoiding negative ones.
- Analyze the employees' readiness for change and define best-practice methods to gain and maintain their support.

B.7.4   Readiness Assessment Data Acquisition

The first step in change readiness assessments is the collection of data and other actionable information. However, in many cases, simply asking employees the following questions can create fear and uncertainty. To help avoid such problems, data should be gathered carefully using a structured change-specific collection and assessment process. Further, leaders must be ready to answer employees' questions when they first learn that change is coming, especially the WIIFM questions. The types of readiness assessment questions that should be answered include the following.

- *What are the various stakeholder groups' perceptions of the organization's readiness for change?*
- *What is the overall degree of individuals' readiness for change in the various stakeholder groups?*
- *What are the various stakeholder groups' assessments of the change itself, and how do they perceive the personal impacts of that change?*

These questions must be included and timed carefully:

- Within the overall CM communications activities, and
- In determining the readiness of the leaders who must communicate change details.

B.8   Anticipating and Managing Obstructions

In addition to defining tasks required to initiate a change, the readiness assessments uncover obstruction risks. Because change obstructions are multifaceted, as are their management methods, this section summarizes many of the concepts change agents and organization leaders will need to use in developing a high degree of CM competency.

B.8.1 The Root Causes of Change Resistance

Change resistance is rooted in the dynamic tension between the *basic employment contract* and *an employee's motivation to work*.

> *The basic employment contract is an agreement wherein the employee trades their time and best efforts to achieve the employer's goals in exchange for the compensation provided by the employer.*

However, that is not an employee's fundamental motivation to work:

> *An employee works to gain the opportunities and resources they need to do the things they really want to do in life.*

Change – especially when frequent, marginally significant, or nebulous – may not be one of their priorities. Although the basic employment contract prevails in conflicts, the more the contract and motivation intersect in a *Venn diagram* sense, the more likely employees are to enthusiastically support change.

There are all manner of reasons for change resistance. Some of the more common ones are:

- A failure of individual leaders to enthusiastically support and drive change initiatives due to their personal values, attitudes, perceptions, and commitments
- A disconnect between the change goals and the organizational structure within which the change is expected to occur
- Characteristics of individuals, including extant skills, learning abilities, attitudes, personality types, personal values, and expectations, and
- Fear of a specific change – or any change for that matter – arising from perceived risks of:
  - Economic loss, including job loss, transfers, and demotions
  - Loss of status
  - Potential social disruptions
  - Inconvenience, and
  - Undefined uncertainties.

B.8.2 Personnel Dispositions Toward Workplace Engagement

McKinsey & Company published a research article in September 2023, *Some Employees Are Destroying Value. Others Are Building It. Do You Know the Difference?* The article suggests six personnel archetypes can help leaders recognize that what works for some people may not work for others. Further, change agents must understand what kinds of employees may or may not be receptive to new initiatives. The research's most disheartening finding, though, is possibly more than 50% of personnel in any organization are not enthusiastic about their employment situation. The implications for change planning, implementation, and institutionalization are that extra targeted efforts are necessary to gain the support and positive engagement of employees who are in the negative categories.

The negative categories include quitters, disruptors, and *the mildly disengaged.* Although not explicitly addressed in McKinsey's research, these categories are related to the recent passive-aggressive *quiet quitting* trend.

- *The Quitters*: McKinsey research found that approximately 10% of the workforce falls into the quitters category. Although they may not be the lowest performers, they are often dissatisfied and less committed, which impacts their performance and leads to quitting for better opportunities.
- *The Disruptors*: Approximately 11% of employees fall into the disruptors category. They are characterized by low satisfaction and commitment levels and have the potential to negatively influence their peers.
- *The Mildly Disengaged*: Comprising about 32% of the workforce, the mildly disengaged exhibit below-average commitment and performance levels. While they may fulfill their job requirements, they lack proactiveness and are generally dissatisfied.

In the generally positive categories are the *double-dippers, reliable and committed,* and *thriving stars*. Disappointingly, the research suggests these combined categories may not represent most employees.

- *The Double-Dippers*: This group accounts for about 5% of employees. They hold multiple jobs simultaneously, often without their employers' knowledge. Double-dippers are dispersed across the satisfaction spectrum, with some engaged while others disengaged.
- *The Reliable and Committed*: The reliable and committed comprise about 38% of the workforce. They are satisfied, committed, and willing to go the extra mile.
- *The Thriving Stars*: Approximately 4% of the workforce falls into the thriving stars category. These are the top talent who bring exceptional value to an organization. They maintain high levels of well-being and performance, and positively impact their teams. However, they are at risk of burnout due to heavy workloads.

B.8.3   Types of Change Resistance

Adverse resistance to change occurs in all manner of ways. Here are some of the more common ones.

- *Resistance to the Change Itself*: People may reject a change because they do not believe in it or believe it is not worth their time, attention, or effort.
- *Resistance to Change Methods*: People may resist change because of inappropriate implementation methods based on coercion, co-optation, and force.
- *Resistance to the Change Agent*: People may direct obstructionist behaviors at the person implementing the change. This often involves personality issues and other less tangible factors. As a result, change agents may find themselves:
  - Having to rigorously defend data, analyses, and solutions against people who are supposed to learn and benefit from them
  - Providing transformational leadership for parts of the organization for which they are not accountable, and
  - Being expected to have all the answers, including solving complex problems on short notice.

Change agents must be prepared to respond to these potential eventualities directly plus indirectly by having their superiors help in the response efforts. Remember, not all resistance is bad. When people resist change, they defend something important that appears to be threatened. Change agents should understand and use such resistance as feedback to improve CM objectives, plans, and implementation methods. Some of the more common overt and covert manifestations of resistance by groups and individuals are:

- Absenteeism
- Unfounded grievances
- The spreading of misinformation
- Reduced productivity, and even
- Sabotage of change activities.

8.8.4 Managing the TANSTAAFL Dilemma

19th Century industrialist and philanthropist John Ruskin is famous for coining the phrase:

> *There ain't no such thing as a free lunch!*

Its acronym, *TANSTAAFL,* is widely recognized in management and technical fields. It means that:

> *For every good thing attempted, there will be adverse side effects.*

Obstructionist behaviors often attempt to stop or alter change initiatives because of anticipated adverse side effects, especially those impacting the obstructionist. This is typical behavior in many highly bureaucratic organizations where maintenance of the *status quo* is culturally valued. However:

> *A leader's job does not require them to find all the reasons why a problem shouldn't, couldn't, or wouldn't be resolved.*

Instead:

> *A leader's job requires them to successfully implement just one workable solution.*

Therefore, a leader is better off focusing on the following principles:

- Adverse side effects vary in significance. However, few are *fatal flaws.*
- With thorough change planning, many adverse side effects can be anticipated and avoided.
- However, when they cannot be avoided, they can at least be mitigated to an acceptable level to permit the change.

B.8.5 Leadership Shortcomings That Lead to Obstructionist Behaviors

Referring back to the concept that *every system is perfectly designed to get the results it gets,* leaders lacking – and therefore failing to practice – requisite management skills often exhibit dysfunctional performance reactions to situations outside their routines. Compounding the problem are their superiors, who allow them to get away with it. Such performance outlooks are based on the following:

- Deflections of accountability, e.g. it's not my job, it's not my fault, etc.
- Aggressive reliance on superficial management styles and overblown performance outcomes to conceal their inabilities, i.e. the Dunning–Kruger Effect, and
- Siloing and defending their turf to limit interactions with and possible intrusions by outsiders who might notice and exploit their managerial shortcomings.

B.8.6 Methods for Overcoming Resistance in Preferred Order

The presence of resistance suggests that something can be done to achieve a better fit among the change, the situation, and the affected people. Therefore, once resistance is detected, change agents must determine the root cause before designing and successfully implementing a corrective action.

There are highly effective positive methods for anticipating and managing resistance, and there are negative ones, too. However, carrots are almost always more effective than sticks. Therefore, the latter should be used sparingly and only as a last resort because of the risk of adverse consequences.

These are some of the more commonly used positive methods to create awareness and support for change. They are vital in the SMS design, implementation, and management processes:

- Proactive, comprehensive communication
- Education, competency-based instruction, and task-specific training
- Involvement and participation at all levels of the organization
- Facilitation and support, and
- Negotiation and agreement.

The negative last-resort methods include:

- Manipulation
- Co-optation, and
- Implicit and explicit coercion.

Despite the earlier point that not all resistance is bad, an effective change agent understands that such behaviors may have damaging underpinnings with potential adverse outcomes. When blatant obstructionists have lost sight of *the basic employment contract* and are acting exclusively within their narrowly defined *motivations to work*, they threaten the success of a change initiative.

Interventions and corrective actions prescribed by human resource policies and procedures are warranted at such times. Further, should the draconian last-resort measures be deemed appropriate, they must be administered rapidly, resolutely, ethically, and legally within established organizational policies and procedures. Despite all the positive efforts to implement change, change agents must be prepared to use the most Machiavellian forms of manipulation, co-optation, and coercion. Further, change agents must never be surprised about whom they must manipulate, co-opt, or coerce.

B.9   Institutionalizing Change

Institutionalizing change is the *refreezing phase* of the Lewin change model. In keeping with Lewin's change model, the book's primary structured refreeze activities are described in Chapter 10's and 11's topics listed below:

- End-of-project and multi-quarter-project reviews
- The annual SMS overall administrative assessments
- Additional PDCA-themed *check* and *act* activities, such as inspections, audits, assessments, and surveys, and
- Restarts of the SMS improvement cycle.

Of course, structure is not enough. There must be unwavering transformational leadership from the board of directors, the chief executive officer, and the executive team committed to the idea that *failure is not an option*. Further, leaders' CM accountabilities for SMS success must include the expectation that personnel throughout the organization will freely welcome the changes over the current situation. As noted in earlier chapters, a critical success element of the SMS is incentivizing accountabilities for all SMS participants to aid in accepting new processes and activities.

B.10   Closing Thoughts on Change Management

After decades of guiding change initiatives, the authors have learned much about CM to transform organizations. To save readers from the frustration of stumbling across these ideas on their own, here are some of the most important lessons the co-authors have learned.

- Change is hard and unwelcome by many in any organization. Why else do so many react to it the way they do?
- Perceptions are reality for the perceivers.
- This idea should be the change agent's mantra:

    People do not commit to decisions because they are sound and rational. They commit to them based on emotions, intuition, trust, hope, and personal values.

- In the end, everyone in the organization must know and understand *what's in it for me*, the *WIIFM* concept. Change agents must understand, encourage, and reward individuals' motivations to change.
- Ultimately, change occurs when people at all levels of an organization take responsibility for the success of the change.
- Spend time building relationships and communicating the benefits of the change among affected stakeholders.
- *Keep it simple, stupid,* is easier said than done. However, adopt the old Nike slogan and *just do it.*
- Prepare yourself for frustration. Change always takes longer and is more complex than was ever imagined. It cannot be designed and installed in an engineering sense.
- Be cautious when trying to change rapidly. As Bob Willard wrote in the *Sustainability Champion's Guidebook*:

    To go fast, you must first go slow.

- It is hard to be patient for results while impatiently working to achieve them. Some days, progress means changing just one person's mind.
- Results matter more than effort.
- Achieving and communicating small successes in the short run eventually builds to larger ones, like a snowball going downhill.
- Recognize and give priority to changing key elements first.
- Change and accountability occur when they are lived, not preached.
- When change agents lose their way and are unsure of how to proceed – which may be often – they need to return to the ideas that ground the change initiative in reality, i.e. why the change benefits the organization and its stakeholders.
- Much of the cynicism about change is due to things starting with great promise and ending with absolutely no difference to people's work lives.
- Too many change initiatives result in cosmetic change. Cosmetic change occurs when planning may be perfect, but results fall far short of expectations.
- It is hard to endure the never-ending challenges from the keepers of the *status quo* and orthodoxy … *really hard!* But, persevere.
- Change agents can work themselves out of their current jobs as organizational competencies increase during the change process. However, new opportunities will open up … *either internally or externally.*
- Although you signed on to the *basic employment contract*, always remember *your motivation to work.*
- Don't take yourself too seriously.

Appendix C

Sustainability Learning and Information Resources

Despite efforts to provide a comprehensive process to create an effective program, this book does not provide sustainability advocates with comprehensive mastery of the management arts and sciences, nor does it provide organizational leaders with working knowledge of sustainability's environmental and social topics. In response to these deficits, though, this appendix examines the need and opportunities for gaining enhanced sustainability knowledge to help individuals and organizations better anticipate and respond to environmental, social, and economic concerns.

First, it summarizes recent needs assessments and commentary on sustainability *knowledge transfer* (KT). It also recommends concepts to be included in the design and delivery of in-house programs to improve their efficacy. Then, it provides representative lists of degree, certificate, and short-course programs primarily available in the United States from universities, professional associations, and public and private sector organizations. It also lists online libraries and other sources of topical sustainability materials.

Including a program or course on a list is neither a recommendation nor an endorsement. Instead, it is simply an acknowledgment that the listed offering was available when writing this book. It is up to readers to determine if the program or course is appropriate to their and their organization's needs. Such considerations include:

- *Is the overall program or course emphasis a good fit for addressing current and future needs?*
- *Is the provider reputable, credible, knowledgeable, and experienced in sustainability?*
- *Is the program and provider accredited by a recognized authority?*
- *What are the academic credentials and organizational leadership experiences of the faculty?*

Sustainability Programs: A Design Guide to Achieving Financial, Social, and Environmental Performance, First Edition. William Borges and John Grosskopf.
© 2025 John Wiley & Sons, Inc. Published 2025 by John Wiley & Sons, Inc.

- *Does the curriculum include relevant courses with practical applications?*
- *What is the delivery modality?*
- *What are the time and effort commitments?*
- *What are the program or course completion success criteria and acknowledgments?*
- *Does the program or course provide good value?*
- *How many students have completed the program or course?*
- *What are the professional profiles of former students, and what career benefits have they accrued since completing a program or course?*

C.1 The Need for Sustainability Education, Instruction, and Training

In its 2022 study, *Closing the Sustainability Skills Gap: Helping Businesses Move From Pledges to Progress*, Microsoft Corporation defined the need for sustainability education, instruction, and training for aspiring and current professionals. The study broadly categorized sustainability jobs as:

- Rapidly emerging sustainability specialties and
- Existing jobs that are expanding in responsibilities to include sustainability concerns.

The study emphasized that as companies create and attempt to fill newly defined sustainability jobs, they will discover a skills gap that presents risks for program success. The following are the kinds of knowledge and skills needed.

- First, deep, specialized knowledge and skills are required in emerging disciplines such as financial and sustainability materiality assessment, carbon accounting, ecosystem valuation, and sustainability performance tracking, analysis, and reporting systems.
- Second, individuals and teams must acquire deep knowledge in specific sustainability subject areas, especially closed-loop life cycle environmental and social issues inherent in value and supply chain management and other administrative and operational activities.
- Third, employees must enhance their technical and change agent skill sets to address sustainability issues in their administration and operations roles.

One of the study's most important findings is:

> Sustainability transformation needs people who can combine specialized sustainability knowledge and skills with other multidisciplinary skill sets.

This involves finding and developing people who can combine *science, technology, engineering*, and *mathematics* (STEM) knowledge with liberal arts disciplines, management, and digital technology. With academic and business emphases on producing and employing conventional professional specialists, finding generalists with broad and deep skill sets and quality experience can be difficult.

This conclusion echoes Joel Makower's July 2021 *GreenBiz* article, *Inside the War for ESG Talent*. In the article, he listed the four qualities recruiters seek and assess in candidates for sustainability leadership positions:

- Business experience
- Adaptive and critical thinking
- A multidisciplinary and systems perspective – *"The nature of ESG and sustainability is as much about influence, education, persuasion, resilience and facilitation,"* and of course
- Passion.

He concluded that ideal candidates are systems-thinking multidisciplinarians who are comfortable with ambiguity, have business experience, can connect the dots, and want to improve the world.

The Microsoft study noted that the growing gap between organizational needs and the number of qualified available people compounds the shortage of people with multidisciplinary attributes. Citing LinkedIn's *Global Green Skills Report 2023*, green jobs grew at an annual rate of 8% between 2015 and 2021, while the talent pool grew at only 6%.

So far, the solution has been *home-growing* the needed talent. Microsoft found that employers hired nearly 70% of their sustainability leaders from within. Further, it was seen that 60% of sustainability team members joined without expertise in the field. These employees have been talented insiders with the core transformational and functional skill sets needed to create change in a company, even though they lacked formal credentials in sustainability. In other words, they have been skilled change agents who have upskilled to accomplish critical sustainability work.

Noting that this approach cannot scale up to meet business or the planet's needs, the Microsoft study recommends these three general actions:

- Organizations must develop a shared understanding of evolving jobs and their requisite sustainability knowledge and skills
- Organizations must move quickly to upskill their workforce through learning initiatives focused on sustainability knowledge and skills, and
- Governments, nongovernmental organizations, and companies must partner to introduce sustainability fluency and science to primary and secondary schools and help higher education institutions strengthen and expand their undergraduate and graduate sustainability programs.

C.2   Types of Sustainability Knowledge Transfer Offerings

As used in this book, *education, instruction,* and *training* in organizational settings have distinct meanings. Accordingly, the general KT term is used here instead of the commonly misused term *training*. It is essential to distinguish between KT terms to prevent unrealistic expectations regarding competency outcomes. When all KT activities are referred to as *training,* when they are not, leaders often expect competencies that their subordinates may not have. In these situations, subordinates are being set up to fail. The following points define these terms to help in-house course designers and facilitators avoid such problems.

- *Education* refers to KT activities that provide general concepts and insights that may be applied sometime in the future to identify, assess, and resolve as-of-yet-undefined problems. Work process or task competencies are not required.
- *Instruction* refers to KT activities that eventually lead to measurable competencies with complex work processes. This definition recognizes that measurable competency cannot be achieved during initial class sessions. It requires specific follow-up, reinforcement, and assessment activities after the initial sessions to achieve and confirm competency.
- *Training* refers to limited-scope KT activities that result in immediate, measurable competency with simple work tasks. Training does not require follow-up and reinforcement, although it can sometimes be beneficial.

While carefully selected degrees, certificate programs, and short courses from reputable institutions and organizations are typically valuable, internally developed programs and courses should conform to performance standards like the *Kirkpatrick Model* to ensure efficacy. The Kirkpatrick Model is a method of evaluating the results of KT programs. It assesses formal and informal KT methods and rates them against four progressing levels of criteria: *reaction, learning, behavior*, and *results*. These evaluation criteria levels are recognizable as *check* steps from the *Plan-Do-Check-Act* cycle.

- Level 1, *reaction*, measures whether learners find the KT engaging, favorable, and relevant to their jobs.
- Level 2, *learning*, formally or informally gauges the KT of each participant to determine if they acquired the intended knowledge, skills, attitude, confidence, and commitment.
- Level 3, *behavior*, determines if the KT has altered participants' behaviors and if they are applying what they learned.

- Level 4, *results*, evaluates the KT results against an organization's performance outcomes, such as *key performance indicators* that were established before learning was initiated.

Unlike decades past, there are scores of KT programs focused on the ever-growing body of sustainability topics. The basic types of KT programs are:

- Academic degrees
- Professional certificate and accreditation programs, and
- Short courses and other resources.

The following sections explain the advantages and disadvantages of each and provide examples for readers to explore.

C.3   Sustainability-themed Master of Business Administration Degree Programs

Sustainability-themed *Master of Business Administration* (MBA) degree programs are ideal for technically oriented professionals and lower- and mid-level managers wishing to advance their careers by transforming their organizations through sustainability principles and practices. Most MBA degree programs offer several areas of concentration, including finance, accounting, organizational development, marketing, and operations management. In recent years, sustainability has been added to the specialization offerings at many institutions. Irrespective of specialization, MBA programs include a standard core business curriculum such as general management, economics, statistics, business law, accounting, finance, operations management, marketing, organizational behavior, ethics, information systems, and international business. Sustainability concentrations add courses such as environmental science and technology, corporate social responsibility, sustainability-focused finance, natural capital accounting, and *environmental, social, and governance* (ESG) reporting. The best programs integrate sustainability topics into core classes.

MBA programs are offered in on-the-ground, hybrid, and online modalities. They range in length from the conventional two-year to the more recent one-year programs. They are provided in several formats, including standard programs with a thesis or capstone project, executive programs for seasoned leaders with or without a thesis or capstone project, and short-term accelerated programs without a thesis or capstone project. The reputation and value of the latter program type should be carefully scrutinized before considering enrollment.

Another caveat is in order. MBA programs prepare students to run parts or all of an organization. Because of constantly evolving operating conditions, they are programs focused on transformational leadership. As such, their focus is not maintenance of the *status quo*. In its most functional definition:

> *Transformational leadership in organizational management means helping enterprises, groups, and individuals progress from a current level of performance to an improved one.*

Transformational leadership requires hands-on knowledge, management skills, and psychological insights to motivate and guide individuals within enterprises and groups to achieve performance improvement goals. These interpersonal leadership abilities cannot be fully developed while a student sits isolated in front of a computer screen for one or two years. To develop leadership skills, students must have person-to-person interactions in challenging problem-solving situations, like the ones experienced in on-the-ground programs and physical workplace environments. Therefore, unless an inexperienced prospective MBA student is engaged full-time in a physical workplace, they should think twice about enrolling in an online program. On the other hand, a seasoned organizational leader may benefit from an online program and should feel free to explore such options.

Here is an alphabetical list of a few sustainability-themed MBA programs in 2024. Again, this is not an endorsement of any program. It is simply an acknowledgment that a sustainability-themed MBA was offered by the institution at the time of the book's publication.

- American University, Kogod School of Business
- Antioch University New England
- Bard College
- Brandeis University, Brandeis International Business School
- Carnegie Mellon University, Tepper School of Business
- Chatham University
- Clark University
- Colorado State University
- Cornell University, Johnson Graduate School of Management
- Duke University, Fuqua School of Business
- Duquesne University, Palumbo-Donahue School of Business
- Harvard Business School
- Illinois Institute of Technology, Stuart School of Business
- Presidio Graduate School
- The University of Vermont, Grossman School of Business
- University of California Berkeley, Haas School of Business
- University of Denver, Daniels College of Business
- University of Michigan, Ross School of Business
- University of North Carolina-Chapel Hill, Kenan–Flagler Business School

- University of Oregon, Charles H. Lundquist College of Business
- University of Southern California, Wrigley Institute for Environment and Sustainability

C.4 Sustainability-themed Professional Certificate and Accreditation Programs

Professional certificate programs are ideal for people who lack the interest, time, or resources to commit to a lengthy degree program yet need to acquire problem-solving knowledge and skills quickly. Further, if they desire or need certification or accreditation to meet organizational or professional requirements. Academic institutions, professional associations, and private and public sector organizations offer certificate programs. Program themes and course topics range from general knowledge to specific competency-focused skill sets. Target students vary from C-suite executives to entry-level technical specialists and everyone in between. Although some are offered through on-the-ground modalities, most are provided online. Time commitments range from a few hours to several days, weeks, or months.

GreenBiz.com published two articles by Trish Kenlon:

- *"Which of These 40 Sustainability Certifications Is Right for You?"*, 3 January 2023, and
- *"Need Sustainability Strategy Training? Look Here"*, 23 May 2023.

The articles described several currently available certificate and professional accreditation programs in the following categories. After confirming they were still available as of the publication date of this book, representative programs from the articles along with ones from the authors' research are listed below in the following categories:

- University certificate programs
- General certifications
- Reporting certifications
- Green building certifications
- Operations in built environments certifications
- Issue- and industry-specific certifications
  - ESG and sustainable finance
  - Governance and risk
  - City and infrastructure, and
  - Other specialties.

C.4.1   University Certificate Programs

- Arizona State University School of Sustainability: *Multiple Sustainability-Themed Certificate Programs Focused on Energy, Economics, Food Systems, and Leadership*
- Cornell University – *Corporate Sustainability Certificate Program*
- Harvard Division of Continuing Education, Professional Development – *How to Build a Sustainable Organization: Challenges, Opportunities, and Strategies*
- New York University's Stern School of Business – *Corporate Sustainability*
- University of California Los Angeles Extension – *Sustainability Certificate*
- University of California San Diego Extended Studies – *Certificate Programs in Sustainability and Behavior Change and Sustainable Business Practices*
- University of Colorado Boulder Leeds School of Business – *Certificate in Environmental, Social and Governance Strategy*
- Stanford Graduate School of Business Executive Education – *Sustainability Strategies: Develop Initiatives to Transform Your Business*

C.4.2   General Certifications

- WholeWorks in partnership with GreenBiz – *Leading the Sustainability Transformation Professional Certification Program*
- The International Society of Sustainability Professionals
  - *Sustainability Excellence Associate (SEA)*
  - *Sustainability Excellence Professional (SEP)*
- The Association of Climate Change Officers (ACCO)
  - *Certified Climate Change Professional (CC-P) Candidate*
  - *Certified Climate Change Professional (CC-P)*

C.4.3   Reporting Certifications

- Global Reporting Initiative (GRI) – *GRI Professional Certification*
- International Financial Reporting Standards (IFRS) Foundation
  - *Sustainability Accounting Standards Board (SASB) Fundamentals of Sustainability Accounting (FSA) Credential*
  - *Integrated Reporting Training Programme Certification*

C.4.4   Green Building Certifications

- The U.S. Green Building Council *Leadership in Energy and Environmental Design* (LEED) Program

- *LEED Green Associate*
- *LEED Accredited Professional with Specialty*
- Green Building Initiative – *Green Globes Emerging Professional, Green Globes Professional, and other certifications*
- International Living Future Institute (IFLI) – *Living Future Accreditation*
- International WELL Building Institute's WELL Building Standards – *WELL Accredited Professional*
- Center for Active Design FITWEL Building Standard – *FITWEL Ambassador Accreditation*
- Social Equity Assessment Method (SEAM) Rating System – *SEAM Accredited Professional (AP)*
- Green Roofs for Healthy Cities Green Roof and Wall Program – *Green Roof Professional Accreditation*

C.4.5 Operations in Built Environments Certifications

- The Association of Energy Engineers (AEE)
  - *Certified Energy Manager*
  - *Certified Business Energy Professional*
  - *Certified Energy Procurement Professional*
  - *Renewable Energy Professional Certification*
  - *Certified Sustainable Development Professional*
- Total Resource Use and Efficiency (TRUE) Rating System – *TRUE Advisor Certification*

C.4.6 ESG and Sustainable Finance Certifications

- Chartered Financial Analyst Institute – *Certificate in ESG Investing*
- Chartered Banker Institute – *Certificate in Green and Sustainable Finance*

C.4.7 Governance and Risk Certifications

- Competent Boards – *ESG Designation Program*
- Global Association of Risk Professionals – *Sustainability and Climate Risk Certificate*

C.4.8 Urban and Infrastructure Certifications

- City Climate Planner – *Urban Greenhouse Gas Inventory Specialist Certification*
- Institute for Sustainable Infrastructure – *Envision Sustainability Professional*

C.4.9   Other Specialty Certifications

- Alliance for Water Stewardship – *AWS Professional, Foundation, Advanced, and Specialist Levels*
- American Center for Life Cycle Assessment – *Life Cycle Assessment Certified Professional*
- Circular Economy Alliance – *Certified Circular Economy Professional*
- Events Industry Council – *Sustainable Event Professional Certificate*
- International Supply Chain Education Alliance – *Certified Supply Chain Analyst*
- ISO 14001 Certifications – *Internal Auditor, Lead Implementer,* and *Lead Auditor*
- Learn Biomimicry – *Biomimicry Practitioner Certificate*
- Product Stewardship Society – *Certified Professional Product Steward*
- The American Institute of Chemical Engineers Institute for Sustainability – *AIChE Credential for Sustainability Professionals (ACSP)*

C.5   Sustainability-themed Short Courses and Other Resources

Short courses and resources are ideal for people who, like those enrolling in certificate programs, must quickly acquire problem-solving knowledge and skills. However, some people do not have the need, interest, or time for professional certification or accreditation. They only need information and skill set enhancements to solve their organizations' current and foreseeable sustainability problems.

As with certificate programs, short courses, and other resources are provided by academic institutions, professional associations, and various private and public sector organizations. Like the certificate programs, course topics range from general knowledge to specific competency-focused skill sets for a wide range of students. Short courses and related resources listed here are typically offered online. Compared to certificate and accreditation programs, time commitments are minimal. While some are free, others cost upwards of a few thousand dollars.

In the May 2023 GreenBiz.com article, *Need Sustainability Strategy Training? Look Here*, Trish Kenlon again listed a few current short courses available at the time of this book's publication. Additional courses and resources are listed in the following categories:

- On-demand online short courses provided by universities and nongovernmental organizations
- Live online short courses provided by private sector organizations, and
- Other sustainability learning resources.

Although not listed here, many university extension programs and community colleges offer sustainability short courses and other KT resources. Check with local institutions for current offerings.

- Duke University Online, Fuqua School of Business – *Impact Measurement and Management for the SDGs (United Nations Sustainable Development Goals)*
- Massachusetts Institute of Technology, Management Executive Education – *Business Sustainability Strategy*
- Principles for Responsible Investment Academy – *Understanding ESG*
- Task Force on Climate-Related Financial Disclosures – *Multiple TCFD Knowledge Hub Online Courses*
- The University of Cambridge Institute for Sustainability Leadership – *Business Sustainability Management*
- United Nations Global Compact Academy – *Sustainable Development Goals Leadership Training Series*:
  - *The Paris Agreement on Climate Change as a Development Agenda*
  - *Integration of the SDGs into National Planning*
  - *Green Marketing Challenge*
  - *Integrated Approaches to Mainstreaming, Acceleration, and Policy Support for the SDGs*
  - *Digital4Sustainability Learning Path*
  - *What is the Net-Zero Standard*
  - *Setting Science-Based Targets to Achieve Net-Zero*
  - *Sustainable Consumption and Production in Africa*
  - *Introduction to risk-informed, conflict-sensitive, and peacebuilding programming*
- University of Colorado via Coursera – *Become a Sustainable Business Change Agent*
- University of Oxford, Said Business School – *Oxford Leading Sustainable Corporations Programme*
- Yale School of Management, Executive Education – *Corporate Sustainability Management: Risk, Profit and Purpose*

- EcoActUs with the Assistance of the Climate Reality Project and Harvard Alumni for Climate and the Environment – *Sustainability Leadership Workshop and Climate Bootcamp*

- OnePointFive Academy – *Sustainability Consulting Accelerator*
- Terra.do – *ESG Fundamentals*
- The Center for Sustainability and Excellence – *Certified Sustainability (ESG) Practitioner Program*

C.5.3   Economics, Finance, and Accounting Learning Resources Provided by Nonprofits

- Accounting for Sustainability Educational Offerings:
  - 18-Month Training Program for Corporate Finance and Accounting Leaders
  - On-the-Ground Workshops on Specific Topics Such as Transition Finance and CSRD
  - Webcasts on Finance, Planning, and Investment Perspectives on Sustainability
- Impact Finance Center
  - Fellowship Program for Sustainability and Finance Professionals in Corporate, Investment and Nonprofit Roles
  - Learning Circles, Investing Accelerators, and Webinars.
- The Ellen MacArthur Foundation – *Circular Economy Introduction*

C.6   Management-themed Learning Resources

The following resources provide instruction in management skillset and other topics important to creating and administering sustainability programs.

C.6.1   Project Management

Project management certificate programs and short courses are available from industry associations, colleges, and universities. The ones listed below provide examples of typical online introductory curricula. Check with local institutions for other offerings. The caveats regarding time commitments, costs, delivery modalities, and course quality apply.

- Project Management Institute, *Project Management Basics – An Official PMI Online Course*
- The University of Arizona, *Introduction to Project Management with CAPM® Exam Prep*
- University of Virginia via Coursera, *Fundamentals of Project Planning, and Management*

Initial *root cause analysis* (RCA) and other performance improvement instruction are widely available from many university extension programs, community colleges, professional organizations, and companies. An online search produces numerous results for downloadable materials, pre-recorded videos, asymmetric online courses, and live online and on-the-ground courses. Time commitments range from a few minutes or hours to a day or more. Costs for these resources range from free to several thousand dollars depending on instruction modality, course content, and provider. Some providers offer continuing education units (CEU). Here are four well-regarded RCA and performance improvement instruction resources:

- American Society for Quality
- Georgia Tech Professional Education
- SixSigma.us
- The Ohio State University, College of Engineering Professional and Distance Education Programs

Numerous for-profit and nonprofit organizations provide KT resources for the various sustainability-related ISO *management system* (MS) standards. Some universities mentioned earlier in this appendix also offer ISO MS courses. As with other offerings, time commitments, costs, delivery modalities, and course quality vary. These are just a few of the more recognizable providers in the United States:

- American Society for Quality
- TÜV SÜD America
- Bureau Veritas North America

C.7 Online Sustainability Libraries and Other Information Resources

The following are not learning programs; instead, they are online libraries and various websites offering sustainability information. While most are free, some have fees to access all content.

- *Environment + Energy Leader's Resource Hub*: This online news organization promotes sustainability and corporate responsibility. It provides newsletters, podcasts, webinars, and a topical database of articles, case studies, and sustainability reports.

- *Global Reporting Initiative*: GRI, an independent international standards organization, helps businesses, governments, and other organizations understand and communicate their impacts on climate change, human rights, and corruption. Its website provides information on reporting standards and their use along with the organization's support services, including education resources, reporting tools, activities of the GRI community, program participants, a goals and targets database, and public policy. Its learning resources are highlighted above in *Section* C.4.3, *Reporting Certifications*.

- *Microsoft Sustainability Learning Center*: Focused on sustainability and technology, this website provides videos, articles, research papers, and other resources developed by Microsoft and industry experts.

- *Trellis, formerly GreenBiz*: Trellis, a leading media and events company, provides business-related sustainability news and topical articles through its online publications and conferences. Its work is dedicated to accelerating the transition to a clean economy, tracking market trends, and advancing the beneficial interplay between business, technology, and sustainability.

- *Ubuntoo Environmental Solutions Platform*: Ubuntoo, a certified *B Corp*, offers free and fee-based AI-enhanced custom database resources and services on sustainability topics.

- *The Ellen MacArthur Foundation*: Committed to creating a circular economy that eliminates waste and pollution, circulates products and materials at their highest value, and regenerates nature, the Foundation offers an extensive library of podcasts, reports, and case studies.

C.7.2   Green Living

- *Recycle Coach*: This website is an education platform that engages residents, businesses, and local governments in recycling activities. It offers free information and fee-based programs focused on local recycling programs and company waste management efforts. For residents and workers, its resources answer the question, "Can this be recycled?" There is also a downloadable smartphone app for participants in municipal and workplace programs.

C.7.3   Environmental Stewardship and Social Responsibility

- *Climate Change Resources*: With an emphasis on climate change, the information resources on this nonprofit's website divide content into *what's happening, its consequences, mitigation*, and *adaptation*. They include reports and papers, as well as media sources.

- *Environmental Health News (EHN)*: Promoting science in public policy, EHN is a publication of nonprofit *Environmental Health Sciences*. Information offerings include news articles, newsletters, and special project reports.
- *Environmental Working Group*: Nonprofit EWG offers an extensive library of environmental and social responsibility consumer guides and research papers.
- *National Environmental Health Association*: For nearly 90 years, NEHA has advanced environmental health science and practice and provided training, education, advocacy, and resources. These resources include books and a searchable database of articles and research project reports on various environmental and social responsibility topics related to health concerns.
- *The National Resources Defense Council*: This nonprofit offers policy-oriented, peer-reviewed data, scientific reports, and issue briefs. It also provides information such as topical *101 guides* and consumer-focused scorecards.
- *United Nations Sustainable Development Publications*: This website is a portal to hundreds of United Nations Sustainability Development Goals (UNSDG) briefs, assessments, research projects, and other reports. The UNSDG learning resources are noted in *Section C.5.1, On-Demand Online Short Courses Provided by Universities and Nongovernmental Organizations*.

C.7.4 Built Environment, Energy, and Technology

- *ENERGY STAR*: This U.S. Environmental Protection Agency site provides product energy efficiency ratings and other information for homes and commercial and industrial facilities.
- *Sustainable Sources*: Akin in some ways to hardcopy trade catalogs from earlier decades, this site is a comprehensive resource for information on notable events and research news, sustainable materials, building codes, and water, energy, and waste management.
- *The U.S. Green Building Council*: The USGBC resources webpage provides access to a searchable database on building design, construction, and operation supporting its green building certification programs. *Section C.4.4, Green Building Certifications* highlights the Council's learning resources.

C.7.5 Sustainability in Business and Government

- *LiveAbout*: Sustainable Businesses – The LiveAbout website dedicates some of its content to sustainable business news and best practices. While the articles are typically rudimentary, they help introduce newcomers to essential sustainability concepts.
- *Office of Federal Sustainability Resources and Guidance*: Specific to federal agencies, this publicly accessible website provides guidance documents on managing

sustainability issues. Many of the documents' principles and practices have applications in the private sector.

- *The Green Infrastructure Wiz*: GIWiz is a repository of EPA-sourced tools and resources designed to support and promote sustainable water management and community planning decisions. The tools are helpful in analyzing problems, defining solutions, calculating design parameters, analyzing costs and benefits, evaluating tradeoffs, and developing education and outreach campaigns.

Appendix D

Special Considerations Regarding ISO Management System Standards

This book's *continuous improvement* (CI) *sustainability management system* (SMS) model can be used to design, implement, and effectively manage a high-performance sustainability program capable of breakthrough performance. Its genesis and development are described in Appendix A.

Applications of the SMS's underlying CI *management system* (MS) model can be used to govern entire organizations, as they have in earlier developmental iterations. However, the model is highly adaptable and can be limited to specific topics, such as the familiar ISO themes of *quality, environmental, energy,* and *health and safety management,* plus more than a dozen others. An MS's focus can be defined broadly, narrowly, or somewhere between. Regarding sustainability, the focus should extend broadly across the full spectrum of financial, social, and environmental topics.

Significant progress has been made in integrating multiple MSs into a single, more efficient one. This is due to innovative organizations' examples and ISO's encouragement through its guidance document, *The Integrated Use of Management System Standards* (IUMSS), 2nd edition, November 2018.

However, there is a caveat. Although the IUMSS has been available for several years, many ISO MS practitioners, consultants, auditors, and registrars remain unfamiliar with its concepts and practical application methods. Likewise, ISO 26000's Guidance on Social Responsibility has also been available for several years. Yet, many practitioners, consultants, auditors, and registrars are comparably unfamiliar and inexperienced. It should go without saying that ISO/UNDP PAS 53002 guidelines, which were introduced in late 2024, are too new to have developed much expertise in their application. Therefore, organizations should exercise caution when seeking support services to certify a sustainability-focused or other integrated MS under ISO standards.

Sustainability Programs: A Design Guide to Achieving Financial, Social, and Environmental Performance, First Edition. William Borges and John Grosskopf.
© 2025 John Wiley & Sons, Inc. Published 2025 by John Wiley & Sons, Inc.

At a minimum, a sustainability-focused MS conforming to ISO requirements should integrate the following standards.

- ISO 9001 – Quality Management
- ISO 14001 – Environmental Management
- ISO 45001 – Occupational Health and Safety Management
- ISO 50001 – Energy Management
- ISO 26000 – Social Responsibility Guidance, and
- ISO/DP PAS 53002 – Guidelines for Contributing to the United Nations Sustainable Development Goals.

Additionally, Social Accountability International's SA8000 Social Certification Standard should be integrated.

Despite any level-of-effort reservations involving creation, administration, operation, or certification, an integrated SMS's operational activities remain conceptually simple within the context of the *Plan-Do-Check-Act* (PDCA) cycle.

- *Plan*:
  - Set organizational policies and create administrative structures.
  - Define, prioritize and shortlist the organization's most pressing needs.
  - Develop strategies, tactics, targets, and capability creation and performance improvement initiatives.
- *Do* and *Check*: Undertake, track, and complete capability creation and performance improvement initiatives to measurably achieve policies, strategies, tactics, and targets, and
- *Act*: Review the initiatives' efficacy, correct problems as necessary, reward successes, and improve the effectiveness of the SMS itself using lessons learned. Then, restart the PDCA cycle.

For those organizations that later choose to secure individual ISO certifications using a comprehensive SMS based on the book's model, some administrative modifications to the initial design may be needed later to conform to ISO requirements. This is not unusual. Changes to achieve and maintain ISO certifications are common, especially when audits reveal deficiencies requiring corrective actions. Again, caution is needed when seeking knowledgeable support services.

As discussed elsewhere in the book, organizations should operate their SMS for at least one PDCA cycle before seeking any ISO certifications unless there are immediate and compelling administration or operations reasons. Once an SMS has produced measurable results and its administration and operations problems have been revealed and corrected, ISO certifications can typically be achieved without undue complications.

Appendix E

Examples of Project Planning Worksheets

Project plans specify how tactical objectives and targets will be successfully achieved on time and within budget by creating new organizational capabilities and correcting performance issues. This appendix provides the four examples of *project management* (PM) worksheets noted in Chapter 10. Worksheets similar to these were used successfully in creating the *United States Marine Corps Base Camp Pendleton Environmental Management System* and in performance improvement initiatives at regional healthcare systems.

These relatively simple worksheets are easy to understand and use in successfully planning, monitoring, and completing a project. They provide a comprehensive, detailed project plan when completed and assembled. In a stepwise project planning sequence, the worksheets are:

- Project Summary
- Task Description
- Project Schedule, and
- Project Budget.

These worksheet examples can be used to produce PM tools with standard office software, such as word processing and spreadsheet applications. If available, performance improvement staff should be involved in the design effort. Further, adaptations of the examples should conform to any established content and formatting standards.

These elementary worksheets are relatively easy for inexperienced project managers to use in either electronic or paper form. Of course, PM software can be used instead. However, beginning project managers typically find sophisticated software applications unnecessarily complex. Their first attempts at project planning should not require the distraction of mastering an unfamiliar software package while learning PM rudiments. Analogous to PM, accountants learn their fundamentals well before ever delving into the intricacies of accounting software systems. Novice project managers deserve the same consideration during their skill set development.

PM support is a prime opportunity to engage the organization's information systems function in developing the SMS. Whatever way project plans are created, they should be *accessible, trackable, amendable, distributable,* and *archivable.* Further, project plans should be linked to the organization's accountability and performance tracking *dashboard, balanced scorecard,* or other reporting systems.

Developing PM competencies throughout the organization is vital to creating an effective SMS. The central role of PM in any continuous improvement management system necessitates at least introductory-level instruction in the worksheets' use in project planning and management. For organizations with training-and-development and performance improvement staff, developing and delivering an internal novice-level PM short course featuring formal follow-up, reinforcement, and competency conformation is a relatively routine effort. As the need for more sophisticated PM skills arises, online courses and offerings from local community colleges and university extension programs are available. Section C.6.1 lists introductory PM course examples for those seeking formal instruction from external sources.

E.1   Project Summary Worksheet

The *Project Summary Worksheet*, shown in Figure E.1, is completed first in the planning process. It summarizes on one page:

- The project name
- Accountable project participants
- A brief description of the project's purpose, including strategic goals, tactical objectives and targets, and the organizational capability that will be created or performance issue that will be resolved
- The project's significance to the organization, including the challenges that must be overcome and the consequences of not completing it on time and within budget
- Intermediate and end-of-project deliverables, and
- Project start, milestone, and completion dates.

E.2   Task Description Worksheet

The second step in the planning process is listing all the major project tasks and then detailing each one on an individual *Task Description Worksheet* like Figure E.2. In PM terms, the list of project tasks is called a *work breakdown*

Project Title	
Project Team	
• Project manager:	
• Project task leaders and contributors:	
• Internal support services and/or external contractors :	
• Assigned sustainability specialty team	
Project Purpose	
• Strategic goals	
• Tactical objectives and targets	
• Describe the organizational capacity the project will create and/or the performance problems it will resolve	
The Consequences of Not Completing the Project On-Time and Within Budget	
Project Deliverables	
• Intermediate deliverables	
• End-of-project deliverables	
Project Schedule	
• Start date	
• Major milestone dates	
• Completion date	

Figure E.1   Project Summary Worksheet.
Source: W. Borges.

Task Number: _____

What is going to be done?	Why is this task necessary?	Where will this task be done & where will resources come from?	When will this task start and finish?	Who will do this task?	How will this task be done?	Check how will performance be checked?
		Work locations:	Start date:	Task leader:	Step-by-step methods:	Programmatic & in-the-moment monitoring activities:
			Mid-task milestones & dates:	Task participants:		Deviation detection criteria:
		Resource providers:	End date:		Deliverables:	Correction prescriptions:

Figure E.2 Task Description Worksheet.
Source: W. Borges.

structure (WBS) that provides a detailed stepwise work process. Please note that as described here, the WBS is a simple list that does not require a separate worksheet; it will be included later on the Project Schedule Worksheet. Because a WBS is a list of tasks in a work process, it should be divided into these three phases underlain by the PDCA cycle:

- Inputs to the work activities – i.e. preparations to do the project work (*Plan*)
- Input-transforming work activities – i.e. successfully completing the project work (*Do* and *Check*), and
- Outputs from and outcomes of the work activities – i.e. confirming the work has accomplished the project task's purpose (*Check* and *Act*).

The accountable project manager and key project participants complete a worksheet for each task in the WBS with the help of the responsible *sustainability specialty team* (SST). Note that the first task in any project plan is continuous *PM*. This task defines how the project will be controlled to achieve the assigned goals, objectives, targets, and capability creation or performance improvement purpose on time and within budget.

When completed, the worksheet describes each task in the *5Ws and 1H + Check* format: *what, why, where, when, who,* and *how,* plus *check.* The following are the key questions answered in the task description worksheets.

- *What is going to be done during this task?*
- *Why is this task necessary?*
- *Where will the task be done, and where will the necessary resources come from?*
- *When will this task start, and when will it be completed? Plus, when are the major in-progress milestones?*
- *Who is the task leader, and who are the other task participants?*
- *How will the task be done, i.e. what is the step-by-step method? Plus, what are the task deliverables?*
- *How will task performance be checked programmatically and in the moment? What are the thresholds for variances and non-conformances? What are the anticipated corrective actions for common variances and nonconformities, non-conformities, or non-compliances to this type of task? (See sidebar.)*

Distinctions Between Variances and Non-conformances, Non-conformities, and Non-compliances

Variances occur when a plan, standard, or regulation is followed, yet the results of the work effort are not as expected. In contrast, non-conformances, non-conformities, and non-compliances happen when a plan, standard, or regulation is not followed. Although such results are typically negative, occasionally they are positive. When the results of such deviations are positive, they should be highlighted in periodic progress and end-of-project reports. However, any positive results stemming from deviations to regulatory requirements raise risk-management concerns.

E.3 Project Schedule Worksheet

Once the task descriptions are complete, a *project schedule worksheet* is used to produce a simple *Gantt Chart,* a standard graphic scheduling tool. Figure E.3 shows a completed project schedule worksheet. The worksheet lists the high-level WBS tasks defined in the Task Description Worksheets and shows their durations and chronological progression. Graphic project schedules:

- Show task predecessors and dependencies
- Reveal task overlaps, and
- Indicate potential resource overloads.

To begin, enter the following on a project schedule worksheet like Figure E.4:

- The project name
- Brief task titles based on the *what* sections of task description worksheets
- The names of the project manager and the respective task leaders, and
- The week-ending dates for each of the 13 weeks in a fiscal quarter.

Then, for each task, fill in a horizontal bar to define a timeline between its start and end dates. In addition to timelines, use symbols, such as asterisks, to show when milestone activities are scheduled. As needed, add a legend to the schedule to define the milestone symbols.

Lastly, adjust both the task descriptions and the schedule when:

- Overlapping tasks overload resources, and
- The draft schedule runs past 13 weeks in cases where the project term is limited to one fiscal quarter.

Figure E.4's project schedule worksheet example must be expanded to accommodate additional tasks and weeks in large, multi-phase projects extending over several fiscal quarters.

E.4 Project Budget Worksheet

Depending on the nature of a project and organizational requirements, a project plan may or may not need a formal budget. In some organizations, smaller performance improvement projects that only use currently available resources will not likely need a budget. However, a budget may be required to assess a project's return on investment more thoroughly or to track sustainability program expenditures. In contrast, a formal budget is always needed for capability creation and other projects that require new hires, outside services, new equipment, new software, extraordinary materials, and additional space or other facility resources.

The *Budget Worksheet* example, Figure E.5, can be enlarged and used in hard-copy form for very simple budgeting. However, for more sophisticated projects,

Project Name: Production Process Waste Reduction Project

Week ending dates	Week 1 3/1/2025	Week 2 3/8/2025	Week 3 3/15/2025	Week 4 3/22/2025	Week 5 3/29/2025	Week 6 4/5/2025	Week 7 4/12/2025	Week 8 4/19/2025	Week 9 4/26/2025	Week 10 5/3/2025	Week 11 5/10/2025	Week 12 5/17/2025	Week 13 5/24/2025
Task 1: Project management / PM: B. Smith													
Task 2: Project kick-off work session / TL: B. Smith													
Task 3: Analyze waste-producing production process / TL: J. Jones													
Task 4: Acquire, test, & commission new equipment / TL: S. Brown													
Task 5: Specify & acquire new production feed stock / TL: P. Garcia													
Task 6: Test run modified production process / TL: A. Miller													
Task 7: Analyze test data & define any OFIs / TL: J. Jones													
Task 8: Implement & test additional improvements / TL: A. Miller													
Task 9: Begin production with the modified process / TL: L. Davis													
Task 10: Periodic & end-of-project performance reports / TL: B. Smith	*	*	*	*	*	*	*	*	*	*	*	*	
Task 11: Formally close the project / TL: B. Smith													

Figure E.3 Example of a Completed Project Schedule Worksheet.

Project Name:												
Week ending dates	Week 1	Week 2	Week 3	Week 4	Week 5	Week 6	Week 7	Week 8	Week 9	Week 10	Week 11	Week 12	Week 13
Task 1: Project management												
Task leader:												
Task 2:												
Task leader:												
Task 3:												
Task leader:												
Task 4:												
Task leader:												
Task 5:												
Task leader:												
Task 6:												
Task leader:												
Task 7:												
Task leader:												
Task 8:												
Task leader:												
Task 9:												
Task leader:												
Task 10:												
Task leader:												

Figure E.4 Project Schedule Worksheet.
Source: W. Borges.

Project Name:

	Task 1			Task 2			Task 3			Task 4			Task 5			Extended task costs
	Unit cost	Quantity	Task cost	Unit cost	Quantity	Task cost	Unit cost	Quantity	Task cost	Unit cost	Quantity	Task cost	Unit cost	Quantity	Task cost	
Labor																
Category A																
Worker 1																
Worker 2																
etc.																
Category B																
Worker 1																
Worker 2																
etc.																
Category C																
Worker 1																
Worker 2																
etc.																
Subtotals																
Services																
Category A																
Category B																
Category C																
etc.																
Subtotals																
Materials & other consumables																
Category A																
Category B																
Category C																
etc.																
Subtotals																
Equipment & software																
Category A																
Category B																
Category C																
etc.																
Subtotals																
Space																
Category A																
Category B																
Category C																
etc.																
Subtotals																
Total costs																

Figure E.5 Project Budget Worksheet.
Source: W. Borges.

it should at the very least be used as a template to format an electronic spread-sheet, enabling project planners to run *what-if* scenarios on a task-by-task basis to arrive at a realistic budget. In addition to providing columns for individual tasks, it breaks tasks down into these cost categories:

- Labor
- Services, both internal and external
- Materials and other consumables
- Equipment and software
- Space and other facility resources.

The initial source of information for cost estimates is Figure E.2's *Task Description Worksheets*. Then, with the assistance of the accounting and purchasing departments and other knowledgeable sources, cost factors can be defined and inserted along with estimated quantities of the requisite resources to calculate expenditures.

Once a project manager completes a draft budget, it must be reviewed by the responsible SST and the SMS champion. However, for larger, more sophisticated, high-budget projects, the SMS oversight group and other responsible departments and persons must ensure that a draft budget can achieve project goals, objectives, and targets with the least cost, effort, and risk. Because it is a draft, the budget will almost certainly be amended before approval. Budget amendments can affect task descriptions, as well as project schedules. Therefore, task descriptions and project schedule worksheets may need amending once budgets are changed.

Appendix F

An Advanced Needs Assessment Method:
Life Cycle Assessment

F.1   Introduction

Early in developing a sustainability management system (SMS), it is usually easy to identify low-hanging fruit, i.e. self-evident opportunities for improvement (OFIs). As discussed in Chapter 6, these OFIs do not necessarily need a formal audit, sustainability materiality assessment, or life cycle assessment (LCA) to be discovered.

However, once the initial OFIs have been successfully exploited, audits, materiality assessments, surveys, and LCAs are the go-to needs assessment methodologies to identify and resolve an organization's other major but less obvious sustainability issues. An LCA is a detailed analysis of environmental and social impacts in all or parts of its value chain. LCAs identify adverse outputs and outcomes within interrelated production and service-delivery systems, as well as OFIs. As such, they provide valuable decision-making information supporting sustainability initiatives.

LCAs are especially valuable for organizations with large-scale and complex value chains producing obscure environmental and social risks and impacts. Because many organizations are familiar with auditing, financial materiality assessment practices, and surveys, there is little need to discuss them in depth outside the *check* function considerations in Chapter 11. In contrast, LCA is a relatively unfamiliar but extremely advantageous body of methods to identy OFIs, hence the need for this appendix.

A thorough LCA is a challenging endeavor. Although some SMS leaders and their organization's technical staff may be tempted to complete a comprehensive LCA in-house early in an SMS's development, it is often inadvisable due to the effort's complexity. However, this is not a suggestion that specialty software be purchased or outside consultants be retained to conduct LCAs. Indeed, simple limited-scope LCAs may be possible and advantageous within an organization's resource capabilities. Nevertheless, organizations should be prudent before undertaking complex LCAs, purchasing specialty software, or retaining consultants.

Sustainability Programs: A Design Guide to Achieving Financial, Social, and Environmental Performance, First Edition. William Borges and John Grosskopf.
© 2025 John Wiley & Sons, Inc. Published 2025 by John Wiley & Sons, Inc.

SMS leaders and participants must know what is involved in conducting an LCA, irrespective of any decisions to secure outside resources. Further, when seeking external assistance, they need to make informed assessment-scoping and contracting decisions that enable them to oversee LCA project progress effectively and achieve desired outcomes. To this end, this appendix provides essential information on the nature of LCA and its fundamental analytical methods.

F.2   What is a Life Cycle Assessment?

An LCA is a complex needs assessment tool that has the potential to comprehensively reveal a product or service's significant adverse environmental, social, and financial risks and impacts throughout its value chain from concept development to beyond the design end-of-life. A *value chain* is:

> *The processes or activities by which a company adds value to a product or service, including design, production, marketing, distribution, and after-sales support.*

The foundational value chain component is its *supply chain*. A supply chain featuring a closed-loop is:

> *A connected system of organizations, activities, information, and resources designed to source, produce, and move goods and services from origination to a final destination, typically from upstream suppliers to downstream end customers and beyond, to include reuse, recovery, and recycling (3R) activities.*

When risks and impacts across a product or service life cycle are revealed in detail, it is possible to prescribe effective avoidance and mitigation measures to remove or reduce often overlooked inefficiencies from entire value chains. These measures can help:

- Increase operational and administrative efficiencies by reducing costs, efforts, and risks, and
- Incorporate environmental and social-benefit attributes into existing and new products and services, enhancing customer loyalty, and attracting new customers to support revenue growth.

The United States Environmental Protection Agency says LCA is:

> *"A technique to assess the environmental aspects and potential impacts associated with a product, process, or service, by:*

- *Compiling an inventory of relevant energy and material inputs and environmental releases*
- *Evaluating the potential environmental impacts associated with identified inputs and releases, and*
- *Interpreting the results to help make more informed decisions."*

The International Organization for Standardization (ISO) also has requirements and guidelines for LCAs with the current ISO 14044:2006 description paraphrased here:

> *LCA addresses the environmental aspects and potential environmental impacts – e.g. use of resources and environmental consequences of releases – throughout a product's life cycle from raw material acquisition through production, use, end-of-life treatment, recycling, and end-of-design-life disposition.*

Although these definitions have evolved from the earlier incomplete *cradle-to-grave* supply chain concept to the complete *closed-loop cradle-to-cradle* sustainability concept, neither addresses all three *people, planet*, and *profit* (3P) sustainability concerns. This is because they limit their focus almost exclusively to environmental issues. Despite these authoritative definitions, by broadening the scope of an LCA to address *people* problems, an organization can use LCA to identify and exploit hidden 3P OFIs lurking throughout its value chain.

F.3   The Life Cycle Perspective

The life cycle perspective provides a closed-loop production and service-delivery framework spanning a product or service's entire value chain. Within this framework, an organization may achieve its sustainability objectives with the least cost, effort, and risk. Figure F.1 is a simplified overview of a life cycle.

The life cycle perspective uses the same *upstream, current midstream*, and *downstream* views as product and service-delivery value and supply chains.

- The *upstream view* refers to the suppliers an organization depends on to produce products and deliver services.
- The *current midstream view* focuses on the organization's internal activities and how they affect the external environments.
- The *downstream view* refers to how the products and services are delivered to and used by customers and how those products and services are dealt with at the end of their design lives.

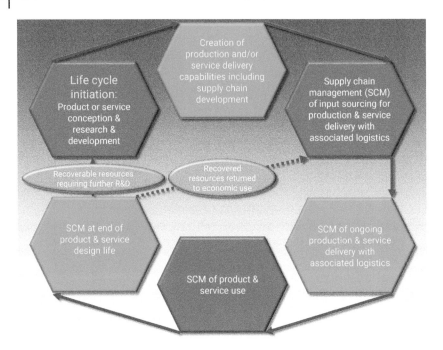

Figure F.1   A Simple View of a Life Cycle.
Source: W. Borges.

The closed-loop dimension added to these value chain and supply chain perspectives enables any generated wastes to be:

- Responsibly returned to productive economic use, or
- Captured and effectively managed in the short term and eventually eradicated through *continuous improvement* (CI) processes in the long term.

F.4   Key Environmental Assessment and Industrial Engineering Ideas in LCA

Focused on value chain management, LCA is a hybrid concept consisting of environmental and social impact assessment disciplines merged with fundamental organizational management and industrial engineering concepts, especially performance improvement. LCAs address two types of significant adverse changes:

- Activities that affect the natural and social environments and
- Activities that can be affected by the natural and social environments.

One of LCA's most crucial industrial engineering ideas is *value engineering*, an operational efficiency concept. Value engineering is:

> *A systematic approach to providing necessary functions in a process at the lowest cost.*

Consistent with the efficiency edict to *achieve objectives with the least cost, effort, and risk*, value engineering strives to maximize function while minimizing cost. So, when inefficiencies are found in a value chain, value engineering promotes substituting materials and methods with less expensive alternatives without sacrificing functionality or measured quality. By extension, in sustainability applications, the LCA concept encourages replacing techniques and materials that cause sustainability-related adverse impacts. From fundamental value-engineering perspectives, a desired outcome of LCA is eliminating nonvalue-added inputs, processes, and outputs.

Value engineering is a central concept underlying CI performance improvement methods, such as those promoted by *W. Edwards Deming, the Japanese Union of Scientists and Engineers (JUSE), Toyota's Lean Production System*, and *Six Sigma*. These inefficiency-eradication methods strive for more salable product and service capacity for every dollar put into production and service delivery. Inefficiency factors are anything that diminishes or prevents financial gains in a value chain, including:

- Regulatory, legal, and other sanctions, including organizational resource diversions associated with avoidable public relations issues, such as greenwashing
- All manner of wastes, e.g. energy, solid, hazardous, liquid, gaseous, financial, personnel, time, and effort
- Mistakes
- Rework
- Delays
- Accidents
- Design flaws
- Product defects
- Cost overruns
- Internal and external health and safety risks, including those associated with defective products;
- Adverse environmental impacts, and
- Social harm.

F.5   Simplicity is a Virtue with LCA

Life cycle assessments can be complicated, even in their simplest forms. Unlike the straightforward process shown earlier in Figure F.1, comprehensive LCAs of entire life cycles are incredibly complex. This is because production and service delivery are not linear processes; they are complex networks. Instead, environmental, social, and financial impacts must be accounted for in each *input, input transformation*, and *output/outcome* step within a life cycle phase. With the multiple life cycle phases shown in Figure F.1, it is evident that there are scores of potential impacts that must be identified and prevented or mitigated. Further, it is easy to fall into the trap of *paralysis by analysis*. Remember, the *KISS principle* is critical to success when conducting an LCA. So, be sure to *keep it simple.*

Analytical simplicity is in keeping with the essential goal of an LCA, which is to merely produce actionable decision-making information. This contrasts with rigorous scientific or engineering analyses that can produce enormous volumes of detail. The results of LCAs are used to:

- Support the organization's sustainability and other policies, goals, objectives, and targets
- Provide a rational basis for operational and administrative sustainability initiatives, and
- Improve overall organizational performance through a sustainability perspective.

F.6   Scoping a Life Cycle Assessment

It is necessary in planning an LCA to:

- Define the product or service life cycle in the phases and steps of its value chain, and
- Scope the assessment in terms of what it will encompass and what it will exclude.

An LCA may be formatted in any number of ways depending on any external requirements as well as internal needs. For example, ISO 14044:2006 specifies four phases:

- Goal and scope definition phase
- Inventory analysis phase
- Impact assessment phase, and
- Interpretation phase.

Formatting conventions aside, the assessment process first involves diagramming a product or service's value chain process map or flowchart featuring one or more closed loops. The closed loops define the actual or ideal cradle-to-cradle relationships in the production or service-delivery value chain. Further, the steps in the value chain diagram must be divided into phases like those shown in Figure F.1. Be aware, though, that some phases may need to be revised or deleted for mature product and service value chains, such as omission of the *research-and-development* and the *initial creation of production capacity and service-delivery capabilities* phases.

Because LCAs are complex, internal and external stakeholders must know what will be included and excluded in the assessment effort by clearly defining the analytical focus, extent, limits, and timing of all work efforts. Therefore, high-level expectations must be set by:

An LCA Example: A Capability Creation Initiative

An LCA was a primary design tool in a capability-creation initiative, a new mining operation in Illinois. To start, a process map was drafted showing the project divided into its life cycle phases. In this case, the phases were design, construction, operations, shutdown, decommissioning, and site restoration. In each phase, environmental, social, operational, and other risks were identified, prioritized, and assessed. Then, specific design-for-environment, as well as risk and impact avoidance and mitigation measures were prescribed in collaboration with engineering, regulatory compliance, and operations teams. Once completed, the LCA's findings and prescriptions were integrated into the draft mine plan. Working with regulatory agencies, the LCA-based final mining plan was completed and implemented after review, modification, and approval. Source: J. Grosskopf

- Stating the required degree of analytical specificity and
- Establishing realistic expectations for the utility of assessment results.

Then, using the closed-loop value chain diagram, scoping decisions can be made considering such questions as the following:

- *Will the LCA be more general in a management planning sense or more detailed in a scientific or engineering sense?*
- *Will the organization's entire catalog of products and services be assessed, or will it be just one or a few?*
- *Will all life cycle phases be considered, or will it be just a few?*
- *Will all or just a few specific inefficiencies be considered, such as the generation of solid, hazardous, energy, aqueous, or gaseous wastes?*
- *Will impacts affecting all stakeholders be assessed, or will the focus be on just the internal or external ones?*

- *Will all or just some temporal considerations be addressed, e.g. past, present, and future?*
- *Will all 3P factors be considered, or just a few?*

F.7   Impact Assessment

The impact assessment techniques used in LCAs are the same as those commonly used in government-mandated environmental impact statements, reports, and assessments, aka EISs, EIRs, and EIAs. The purpose of these methods in an LCA is to:

- Identify and characterize the risks, occurrences, and consequences of significant adverse environmental and social impacts, as well as financial ones, and
- Prescribe obviation and mitigation measures.

In doing so, the analytical methods identify sustainability-related impact-causing inefficiencies throughout a value chain's closed-loop life cycle. Once identified, these inefficiencies can be prevented, eliminated, or minimized to achieve organizational goals with the least cost, effort, and risk.

F.7.1   What Impacts are in an LCA … And What They Should Not Be

Organizations should be analytically conservative in defining what a sustainability-related impact is … *and what it is not*. To that end, the following definition is recommended, the rationale for which is discussed later in this section:

> *An impact is a significant <u>adverse</u> change in the natural or human environments.*

The determination of *significance thresholds* can be challenging and must be determined individually. This can be on an impact-by-impact basis or a general basis. The setting of situational significance thresholds is a clear example of the book's mantra, *"one-size does not fit all."* The critical question is:

> *How bad does an impact have to be regarding severity, including cost factors, frequency, and exposure before it is considered internally, legally, technically, and politically significant?*

Note that what is acceptable to an organizational leader or lawyer often differs from a technical practitioner's perspective. Further, leaders, lawyers, and practitioners can be out of line with regulator and activist views of significance. So, remember:

A significance threshold might both be legally and technically correct. However, it might still be incorrect if it is not perceptually correct to the most critical external stakeholders.

This strict definition of adverse impacts has a vital risk management *fail-safe* quality. It precludes assessments of *positive* or *beneficial* impacts. By doing so, attention is focused exclusively on eliminating or mitigating the adverse effects of products and services occurring throughout their life cycles. Further, this definition precludes *greenwashing*, where intentional or unintentional attempts are made to *balance, buffer, neutralize*, or otherwise *excuse* adverse impacts with beneficial ones. As a result, this fail-safe definition is instrumental in preventing corporate leaders, marketing and public relations staff, and others from embarrassing an organization – *or much worse* – through greenwashed misuses of LCA findings.

Despite the fail-safe quality of this strict impact definition, some readers may still question the need for an exclusive focus on adversities in LCAs. After all, the other three non-LCA needs assessment methods described in the book – *SWOT analyses, materiality assessments*, and *audits* – consider beneficial organizational, production, and service attributes.

Because other needs assessment methods allow positivity, it might be reasonably assumed that LCA methods should also. ISO 14001's environmental aspects reinforce this assumption with an impact definition that includes benefits assessment.

Notwithstanding permissive impact definitions, LCAs should not assess beneficial impacts. This is because LCA methods have been adapted from analytically conservative environmental impact assessment and industrial engineering process improvement techniques that concentrate on discovering and remedying adversities. In the case of industrial engineering, process improvement analyses focus on relentlessly identifying and eradicating *inefficiencies* in the value chain, i.e. inabilities to achieve objectives with the least cost, effort, and risk. In simple terms, *they focus on getting the junk out of products and service value chains.*

The central idea behind this emphasis on adversity is that all work processes are initially designed to produce beneficial results. However, many processes have inefficiencies inadvertently designed in. Further, over time, all processes accumulate inefficiencies through variances and nonconformances to standard operating procedures, resulting in unnecessarily weak financial performance and adverse impacts in the natural and social environments. Such inefficiencies limit the achievement of the original process goals.

In conclusion, as a specialty performance improvement methodology conforming to the value-engineering concept:

LCA's purpose is to identify and ultimately eradicate 3P inefficiencies, leaving only a beneficial closed-loop value chain.

F.7.2   Adverse Impact Characteristics

As noted above, causes of adverse impacts are those activities that:

- Affect the natural and social environments and
- Activities that can be affected by the natural and social environments.

Impact categories include:

- Short- and long-term adverse changes
- Direct and indirect adverse changes
- Acute, chronic, and cumulative adverse changes
- Site-specific, local, regional, national, and global adverse changes, and
- Avoidable and unavoidable adverse changes.

Production and service-delivery concerns include:

- Land use activities, including facility design, construction, operation, and decommissioning
- Resource extraction, processing, and use
- Water sourcing, treatment, and use
- Energy production and use
- Chemical, biological agent, ionizing radiation, and other hazardous material use
- Controlled and uncontrolled releases of gaseous, aqueous, solid, hazardous, and energy wastes
- Health and safety
- Demands on public infrastructure and services
- Transportation and other logistics systems, and
- Waste management.

Physical, biotic, and human environment assessments typically are sorted into these categories:

- Physical impacts involving:
  - Atmospheric conditions
  - Geologic and physiographic resources
  - Aquatic and marine resources
- Biotic impacts involving:
  - Ecological systems
  - Botanical resources
  - Zoological resources
- Human impacts defined by the *Global Report Initiative* (GRI) and other social responsibility standards, including:

- Human rights indicators
- Labor practices and decent work indicators
- Social justice inside and outside of an organization
- Small- and large-scale health and safety indicators
- Societal quality of life indicators
- Product responsibility indicators
- Economic indicators
- Protection of cultural resources, and
- Esthetics.

F.7.3 Impact Assessment Steps

The ideal condition for conducting impact assessments is having all necessary information readily available. *However, there are no ideal conditions.* There will never be enough resources to produce a technically perfect impact analysis. However, as noted above, technical perfection is not necessary because impact analysis aims to provide decision-makers with actionable management information. It is not intended to be an exhaustive scientific research tome. To treat impact analysis as anything more than a way to produce actionable decision-making information risks falling into the *paralysis-by-analysis trap* whereby simple, effective solutions are overlooked, and valuable resources are squandered.

For each of the steps in the life cycle phases shown in this appendix's Figure F.1, a comprehensive LCA attempts to answer the following questions considering the various adverse impact characteristics noted above.

- Effects of Adverse Changes on the Environment: *Have there been, are there currently, or will there be any adverse changes to the natural or human environments resulting directly or indirectly from a work activity, product, service, or an associated release of wastes to the environment? If so, what are the adverse changes?*

- Effects of Environmental Conditions on an Organization's Activities: *Have there been, are there currently, or will there be any adverse risks to or actual effects on work activities, products, services, or associated releases of wastes and emissions to the environment attributable to extant natural or human environmental conditions? If so, what are the risks and adverse effects?*

Once the adverse risks and effects have been identified, their respective causal factors and degrees of adversity are detailed following these steps.

- *Adverse Risk and Change Factors*: Define the risk or change's contributing factors in terms of what caused, causes, or will cause them.

- *Determination of Adversity*: Explain why the *severity, frequency*, and *exposure* factors of the risk or change are adverse in both qualitative and, when possible, quantitative terms.

Because an LCA can identify scores of OFIs that must be prioritized, the *Failure Modes and Effects Analysis* (FMEA) method can rank and short-list adverse impacts and risks. (Refer back to Figure 7.6 in Chapter 7.) Remember the cautionary notes discussed earlier in Chapter 7's Section 7.8 about determining significance thresholds when setting actionable FMEA cut-off values.

F.7.4 Prescription of Obviation and Mitigation Measures

The last step in an impact assessment is the definition of obviation and mitigation measures. From a value-engineering perspective, designing and successfully implementing obviation measures to prevent impacts and risks is preferable to mitigation measures. The most effective ways to do this are alterations to the production or service-delivery process steps and inputs. The focus of such efforts is:

- Deletion or modification of problematic process steps and
- Elimination or substitution of adverse-impact-causing materials.

Although the preferred approach is the absolute avoidance of a negative impact, it is not always possible. So, the next best thing is to mitigate risk and impact adversity to an acceptable level. The concerns about setting appropriate significance thresholds discussed above apply here, too.

Prescriptions for acceptable obviation and mitigation measures are developed by asking questions like these for each significant impact:

- *Can the causal factors leading to an adverse impact be removed from a production or service-delivery process, thereby eliminating the impact's risk or occurrence?*
- *If the adverse risk or impact cannot be eliminated, how can severity – including costs – frequency, and exposure factors in a production or service-delivery process be modified to reduce it below its significance threshold?*

F.8 Inclusion of LCA Findings and Recommendations in an SMS

Life cycle assessments help organizations achieve sustainability's promise and full value by identifying and exploiting important new and not-so-obvious OFIs that other needs assessment methods often miss. The process for including these new OFIs in an SMS is described in Chapter 7's Section 7.9 *Aggregating, Reviewing, and Approving the SWOT Analysis Findings.*

Glossary

5Ws & 1H + Check Format This term is shorthand for a format used to define work process activities. It answers the following questions. *What is going to be done? Why is this necessary? Where will it be done, and where will resources come from? When will this start and finish? Who is involved? How will it be done? And how will it be checked?*

Accountability As used in this book, an accountability is an assigned responsibility for the accomplishment of a goal, target, initiative, project, or task. The performance of such responsibilities is tracked, reported, and evaluated. A key consideration of evaluations is determining eligibility for available incentive rewards. See *Performance Incentives.*

Adult Learning Theory Adult learning theory enhances teaching methods and experiences for adult learners. It focuses on providing students with an understanding of why they need to learn something, along with hands-on experiences so they can address that need.

Basic Employment Contract The basic employment contract is an agreement wherein the employee trades their time and best efforts to achieve the employer's goals in exchange for the compensation provided by the employer. Contrast with *Employee's Motivation to Work.*

Benchmarking Benchmarking consists of various needs assessment efforts by an organization to compare its products, services, and processes against those of leading organizations. These efforts identify opportunities for improvement. See *Opportunities for Improvement.*

Best Management Practice Best management practices (BMPs) were originally defined as methods determined to be the most effective and practical means of preventing or reducing nonpoint source pollution to help achieve water quality goals. Since then, the term has acquired broader applications in management and other technical fields. Now, the term refers to any set of proven management methods based on current industry standards

Sustainability Programs: A Design Guide to Achieving Financial, Social, and Environmental Performance, First Edition. William Borges and John Grosskopf.
© 2025 John Wiley & Sons, Inc. Published 2025 by John Wiley & Sons, Inc.

and practices. They involve a wide range of tools, from planning and design to performance improvement.

Breakthrough Performance A breakthrough is achieved by radically altering the underlying structure of a system, process, or product to achieve a dramatic level of performance improvement. See *Capability Creation Initiatives* and *Performance Improvement Initiatives*.

Capability Creation Initiatives As used in this book, a capability creation initiative is typically a project to provide new resources to accomplish a cascading policy, strategic goal, tactical objective, and target. Capability creation initiatives typically involve capital expenditures and personnel development efforts. They differ from performance improvement initiatives, where work processes are modified to better use existing resources.

Catch-as-Catch-Can This phrase describes a situation where people must improvise or do what they can with limited means.

Change Management Change management (CM), a subset of organizational development, is a collective term for various methods to prepare, support, and help individuals, groups, and organizations change their work activities. At its simplest, CM involves assisting people in modifying how they are currently working to produce better results. See *Organizational Development*.

Chief Executive, Chief Executive Officer (CEO), Executive Director These top-level executives formulate policies and plans and provide overall leadership at for-profit companies, nonprofits, and public-sector organizations within guidelines set by a board of directors or similar governing body. They control administrative and operational activities at the highest management level with the help of subordinate executives and staff managers.

Closed Loop Value Chain A closed-loop value chain consists of all the activities by which a company adds value during the entire life cycle of a product, service, or work process. The general activities include design, material sourcing, production, marketing, distribution, consumption, and especially end-of-service life disposition of materials through reverse logistics processes. See *Life Cycle*, *Reverse Logistics, Supply Chain,* and *Value Chain*.

Conformance versus Compliance Conformance generally refers to an organization meeting an industry standard, another external performance requirement, or an internal process such as a standard operating procedure or project plan. In contrast, compliance generally refers to an organization fulfilling its legal, contractual, and other requirements. See *Nonconformances, Nonconformities, and Noncompliances*, and *Variances*.

Constraints A constraint in a systems environment is a limitation or impediment preventing a successful outcome of a planned or intended action, activity, or initiative.

Continuous Improvement Continuous improvement refers to recurring activities that enhance or improve the performance of an organization's products, services, and processes in incremental or breakthrough ways.

Continuous Improvement Management System A continuous improvement management system is structured on an iterative schedule to produce improvements to products, services, and processes through incremental and breakthrough improvements. See *Management System* and *Sustainability Management System.*

Continuous Improvement Methodologies This is a large body of analytical and problem-solving methods focused on incrementally and iteratively enhancing product and service value while identifying and eliminating wastes in their many forms. Established methods are *Lean Manufacturing, Six Sigma,* and *Total Quality Management,* all of which stress systematic and data-driven approaches.

Corporate Social Responsibility Corporate social responsibility (CSR) is a business concept that helps a company be socially accountable to itself and its many stakeholders. Its four elements are environmental, philanthropic, ethical, and economic responsibility. CSR is now an integral component of organizational sustainability. As such, it extends to public and nonprofit organizations.

Diplomatic Assertiveness Diplomatically assertive people state their opinions while still being respectful of others. In contrast, aggressive people attack or ignore others' opinions in favor of their own. Passive people do not state their opinions at all. These are important distinctions. Diplomatic assertiveness should be taught and encouraged in change management planning, implementation, and institutionalizing activities. The other two should be discouraged.

Dunning-Kruger Effect In simple terms, the Dunning–Kruger effect occurs when a person's lack of knowledge and skills in a particular area causes them to overestimate their competence. By contrast, this effect also causes those who excel in a given area to think the task is simple for everyone and, therefore, underestimate their relative abilities.

Efficiency/Inefficiency The term efficiency refers to the achievement of objectives with least cost, effort, and risk. Inefficiency is the failure to achieve objectives with least cost, effort, and risk. See *Sustainability Efficiency.*

Employee's Motivation to Work An employee works to gain the opportunities and resources they need to do the things they really want to do in life. Contrast with the *Basic Employment Contract.*

Environmental, Social, and Governance (ESG) ESG refers to a collection of organizational performance evaluation criteria used to assess and report the robustness of governance mechanisms and their ability to effectively manage environmental and social impacts.

Fail-Safe Fail-safe is a concept wherein a system, plan, or design prevents adverse events from occurring.

Fiduciary Responsibility This is a legal responsibility for duties of care, loyalty, good faith, confidentiality, and more when serving the best interests of a beneficiary. In the case of for-profit businesses, beneficiary refers specifically to the responsibilities an organization has to its equity holders and generally to its other stakeholders.

Goodwill in Business Goodwill is an accounting term that refers to the value a company derives from its brand, customer base, and reputation associated with its intellectual property. Goodwill is a long-term intangible asset that generates value for a company over the long term. Sustainability issues can be significant factors in determining the monetary value of goodwill.

Greenwashing Greenwashing is the act of making false or misleading statements about the environmental benefits of a product, service, or practice. It falsely implies that a company or its products, services, and practices are environmentally and socially responsible. See *Window Dressing*.

Intuitive Decision-Making Intuitive decision-making is based without evidence on personal knowledge, perception, belief, instinct, emotion, or personality traits. See *Rational Decision-Making*.

Key Performance Indicators Key performance indicators (KPIs) are quantifiable measurements used to gauge a company's overall short- and long-term performance. KPIs specifically help determine an organization's strategic, financial, and operational achievements, especially compared to others within the same sector.

Kirkpatrick Model of Training Evaluation This standard knowledge transfer evaluation method focuses on four topics:

- Level 1: Reaction – The degree to which participants find the training favorable, engaging, and relevant to their jobs
- Level 2: Learning – The degree to which participants acquire the intended knowledge, skills, attitude, confidence, and commitment based on their participation in the training
- Level 3: Behavior – The degree to which participants apply what they learned during training when they are back on the job
- Level 4: Results – The degree to which targeted outcomes occur as a result of the training and the support and accountability package.

Although the Kirkpatrick model uses the term "training" for all knowledge transfer activities, note the distinctions between education, instruction, and training in *Knowledge Transfer*.

Knowledge Transfer Knowledge transfer (KT) is a body of approaches to the sharing of information or ideas. The three types of formal KT should never be confused. Distinctions must be made between general *education*, competency-focused *instruction*, and task *training* to prevent unrealistic outcome expectations by those who mandate and provide KT, as well as those who receive it.

- Education – Education refers to KT, where general information is disseminated to raise awareness that may be useful later. No process- or task-related competency is expected.
- Instruction – Instruction has the same goal as training, i.e. competency. It is used when competency must be measurably achieved with complex work processes. Instruction requires one or more KT sessions with follow-up and reinforcement activities before measuring and certifying competency.
- Training – The goal of training is competency. However, this frequently misused term should be restricted to KT activities involving only the simplest tasks. Training activities should be accomplished in one session during which the students can measurably demonstrate competency by the completion. Follow-up or reinforcement is not usually necessary, but can be useful in some situations.

Life Cycle There are two definitions of the term "life cycle" applicable to sustainability: the organizational definition and the value chain definition. The broad organizational life cycle definition has four stages: startup, growth, maturity, and decline. Organizations constantly risk slipping into the decline stage from any of the earlier three. With their emphases on risk management, cost and expense reduction, revenue enhancement through innovation, and greenwash-free transparency, effective sustainability programs provide additional management capabilities to avoid decline. Life cycles with this definition are not necessarily closed loops. The more focused value chain definition stresses product design, manufacturing, distribution, use, and end-of-life disposition. The definition can also be applied to administration and service delivery processes with modifications. Life cycles with this definition are closed loops, wherein end-of-life disposition involves returning spent materials to productive economic use. See *Life Cycle Assessment* and *Reduce, Reuse, and Recycling (3R) Concepts*.

Life Cycle Assessment A life cycle assessment (LCA) is a detailed analysis of environmental and social impacts in all or parts of a value chain. LCAs identify adverse outputs and outcomes within interrelated production and service-delivery systems, as well as opportunities for improvement. As such, they provide valuable decision-making information supporting sustainability initiatives. See *Life Cycle* and *Needs Assessment* and *Value Chain*.

Management System A management system is a collection of interrelated, interdependent, and interacting functions that establish policies, structures, processes, and activities to achieve goals, objectives, and targets that respond to an organization's needs. The most common type of management system is the overarching one used to control an entire organization's activities. See *Continuous Improvement Management System* and *Sustainability Management System*.

Materiality Assessment In business management, the term materiality refers to the issues most important to an organization. Materiality assessment is a standard needs assessment term favored by large publicly traded corporations. Regarding sustainability programs, it is essential to distinguish between the terms *financial materiality* and *sustainability materiality*. In financial reporting, information is material if its omission or misstatement could influence the economic decisions based on financial statements. In contrast, sustainability materiality refers to issues that may have significant repercussions; however, formal monetary thresholds have not yet been defined to determine their financial materiality. As with any needs assessment, some companies retain external consultants while others manage the process themselves. See *Needs Assessment*.

Monitor-Detect-Correct Method The monitor-detect-correct (MDC) method elaborates the "Check" and "Act" activities in the PDCA cycle and the 5Ws & 1H + Check format. It divides these two activities into three parts: periodic and in-the-moment monitoring of work activities; detecting and immediately reporting variances and nonconformances to performance standards; and immediate definition and implementation of corrective actions.

Needs Assessment A needs assessment is a process for determining the gaps between current and desired outcomes within a particular process or system. A need is an opportunity for improvement. Common sustainability needs assessment methods are surveys, materiality assessments, life cycle assessments, and SWOT analyses. See *Life Cycle Assessment*, *Materiality Assessment, Opportunities-and-Constraints Assessment, Opportunities for Improvement,* and *SWOT Analysis*.

Nonconformances, Nonconformities, and Noncompliances While there are technical and legal distinctions between the terms, they functionally describe the same types of situations. Nonconformances, nonconformities, or noncompliances occur when requirements – including regulations – are not met or standards or plans are not followed. Note that outside of strict interpretations, these three terms are often used interchangeably. The broadest, nonconformance, is used throughout the book as shorthand for any of the three. When a work effort's results – and often the work effort itself – differ from what is mandated, planned, or expected, the output or outcome is typically negative. When the results are negative, actions must be taken to control or correct them and deal with the consequences. However, with the exception of regulatory noncompliances, outputs and outcomes can be positive. When they are, they should be highlighted as lessons learned in the periodic project-progress reports, the end-of-project closure reports, and management systems reviews. See *Conformance versus Compliance* and *Variances*.

Nongovernmental Organizations NGOs are nonprofit citizens groups organized on local, national, or international levels to address public-good issues.

Organizational Culture Organizational culture is the formal and informal sets of values, beliefs, attitudes, systems, and rules that influence employee behavior and performance within an organization.

Opportunities-and-Constraints Assessment Complementing broad-scope needs assessment methods, this highly focused management assessment is often an informal consideration or can also be a formal analysis of:

- Opportunities for specific performance improvements and
- Limitations that restrict or prevent exploitation of those opportunities. Such assessments typically consider the availability of resources and returns on investment. See *Needs Assessment* and *Opportunities for Improvement*.

Opportunities for Improvement Opportunities for improvement (OFIs) arise when strengths, weaknesses, opportunities, and threats are identified in external and internal operating environments, structures, and processes. Various audits, systems reviews, and other needs assessments identify OFIs that can benefit an organization if exploited or harm it if not addressed. See *Needs Assessment*.

Organizational Development Organizational development (OD) involves planning and implementing systematic changes in employee work activities to produce overall organizational growth. Typical OD functions include organizational effectiveness, communication process improvement, change management, training and development, product and service improvement, and financial performance improvement. OD expertise is often found in human resource (HR) departments, although the activity is far broader in scope than HR. See *Change Management* and *Training and Development*.

Organizational Transparency Organizational transparency is sharing information regarding the organization's policies, plans, administration, and operations with its stakeholders to promote collaboration and build clarity, trust, and accountability. The concept is the basis for sustainability and ESG reporting standards and requirements.

Organizational Units, Functions, and Departments Units, functions, and departments are organizational subdivisions. As used in this book, "unit" refers to strategic business units (SBU) and their analogs in the public sector and nonprofits. An SBU is a profit center focused on product and service offerings in specific market segments. They can be standalone companies under larger entities, operating divisions, or branches. The term "function" in the book refers to activities spanning several organizational departments. One example is the supply chain management function, which minimally includes operations planning, purchasing, inventory control, production, and logistics departments. "Department" refers to specialty activities organized as cost or profit centers within an SBU. Departments can be subdivided further into sections and teams.

Pareto Principle The Pareto Principle – aka *the 80/20 rule* and *the vital few* – asserts that 80% of consequences come from 20% of the causes. Named after early 20th-century economist Vilfredo Pareto, it was developed further in the context of quality control and improvement during the 20th century by management consultant Joseph M. Juran.

Performance Improvement Initiatives Based on the value engineering concept, performance improvement – also known as process improvement – is a body of efficiency methods involving the: measurement of work process activities; identification of task-specific failures to add value to the work process with the least cost, effort and risk; and modification of the work process to eliminate those failures and inefficiencies using best practices. Standard performance improvement methods include *Lean Manufacturing, Six Sigma*, and *Total Quality Management*. See *Capability Creation Initiatives* and *Value Engineering*.

Performance Incentives Performance incentives are rewards used to motivate positive behaviors. Incentive programs are designed to attract, retain, and especially engage personnel. Common incentives include:

- Monetary bonuses
- Promotions
- Salary raises
- Professional development opportunities
- Additional paid time off, and
- Nonmonetary awards and other formal and informal types of recognition and expressions of appreciation.

Plan-Do-Check-Act Continuous Improvement Cycle The Plan-Do-Check-Act (PDCA) cycle is an iterative management concept that requires focused and systematic engagement to create, operate, monitor, and improve work processes and their resulting outputs and outcomes. It is the foundational concept underlying continuous improvement methods and systems.

Primary Functions of Organizational Management
- Planning – The definition of organizational goals, actions, and resources.
- Organizing – The designation of organizational structures, administration and operations processes, lines of authority, resource allocations, and overall coordination processes. See *Organizational Development*.
- Controlling – The process of monitoring activities, measuring performance, comparing results to objectives, and making modifications and corrections when needed. See *Monitor-Detect-Correct Method*.
- Leading – Depending on both the psychology and engineering schools of change management, leading is the function that helps groups and their individuals move beyond current levels of performance to better ones in keeping with the efficiency concept, i.e. achieve objectives with the least cost, effort, and risk. See *Change Management* and Transformational Leadership.

Quick-Response Corrective Actions *Point-Kaizen events, tiger- and red-team interventions, small-tests-of-change,* and *quick-hits* are conducted to find solutions quickly. By bringing together a group of focused experts, problem-solving can be accelerated to resolve inefficiencies in a work process in minutes, hours – or, at the very most – a day or two.

Rational Decision-Making Rational decision-making uses objective knowledge and logic to identify and analyze a problem. In doing so, it identifies options, considers all relationships, and selects the best alternative. See *Intuitive Decision-Making*.

Reduce, Reuse, and Recycling (3R) Concepts This is a set of circular economy waste management strategies and tactics first developed by the United States Environmental Protection Agency. They are listed in the preferred order of application to minimize the use of new materials and energy by returning entire products or their constituent materials to productive economic use. See *Life Cycle* and *Reverse Logistics*.

Return on Investment Return-on-investment (ROI) calculations are frequently used to develop objective support for a project proposal. A simple ROI formula is net income divided by the total cost of the investment. In the case of sustainability, ROI calculations are expanded to identify, quantify, and consider environmental and social costs and benefits along with the financial effects of project investments.

Reverse Logistics Reverse logistics is a body of disciplines and processes that close value chain loops by returning end-of-design-life and other materials to productive economic use. It involves returning, repairing/refurbishing/remanufacturing, repackaging, reselling, reusing, and recycling. See *Reduce, Reuse, and Recycling (3R) Concepts*.

Shared Governance Shared governance is a form of participative management that modifies conventional top-down organizational control. It is commonly used in healthcare and education. It involves the creation of participation structures and processes for partnership, equity, accountability, and work-process ownership with an organization's stakeholders. It is a highly effective organizational development and change management concept in proactive organizations.

SMART Concept This acronym is shorthand for a set of program and project management criteria used to evaluate initiative goals, i.e. specific, measurable, achievable, relevant, and time-constrained.

Strategic Goals Strategic goals provide an organization with broad thematic directions for future long-term initiatives and activities. Contrast this with the more detailed short-term tactical objectives and targets. See *Tactical Objectives and Targets.*

Sustainability The United Nations' high-level definition is: "Meeting the needs of the present without compromising the ability of future generations to meet their own needs." In organizational management terms, it can be defined this way: *Throughout a product or service's entire closed-loop life cycle, sustainability is how an organization creates value by maximizing its activities' positive social, environmental, and economic effects while eliminating or minimizing their adverse effects.*

Supply Chain A connected system of organizations, activities, information, and resources designed to source, produce, and move goods and services from origination to a final destination, typically from a supplier to an end customer and beyond, including reuse, recovery, and recycling (3R) activities in closed-loop supply chains. They are a fundamental component of value chains. *See Value Chain.*

Sustainability Efficiency This term refers to achieving the organization's people, planet, and profit goals with the least cost, effort, and risk throughout the closed-loop life cycles of its products and services. See *Efficiency/ Inefficiency.*

Sustainability Management System As the structured implementation activities of a sustainability program, a sustainability management system (SMS) is a collection of interrelated, interdependent, and interacting functions that establish policies, structures, processes, and activities to achieve goals, objectives, and targets that respond to an organization's people, planet and profit needs. See *Management System, Continuous Improvement Management System,* and *Sustainability Program.*

Sustainability Program A sustainability program is an organization's general body of measures and activities that drive people, planet, and profit performance improvement results. A sustainability management system is an essential component of a program, providing its administrative and operational elements. *See Sustainability Management System.*

SWOT Analysis This situational analytical technique enables an organization to identify its strengths, weaknesses, opportunities, and threats in the early stages of organizational planning. The method can be employed broadly for enterprise-wide planning, and it can also be focused narrowly on programs such as sustainability and individual work processes. Although it can be highly detailed and objective, it is often used collaboratively to define subjective group consensuses.

Systematic versus Systemic Something described as systematic uses or follows a system, while something described as systemic is part of or is embedded in the system itself.

Systems Thinking Systems thinking is a holistic approach to analysis and problem-solving that focuses on how a system's constituent parts interrelate and how systems work over time and within the context of larger systems.

Tactical Objectives and Targets Derived from an organization's general strategic goals, tactical objectives provide specificity in setting qualitative and quantitative performance requirements for designated units, functions, and departments. See *Strategic Goals.*

The Prime Directive of Finance A prime directive is a requirement that establishes the overriding control over a course of action. In the case of organizational management, the prime directive of finance is *no money, no mission.* It is the pre-eminent consideration in sustainability's people, planet, and profit concept. An organization cannot fulfill its social and environmental responsibilities without adequate financial performance. See *Triple Bottom Line.*

The WIIFM Concept WIIFM stands for "What's In It For Me?" It is a value proposition concept based on the premise that people are more likely to embrace change if they understand how it benefits them personally.

Training and Development Training and development (T&D) – a subset of organizational development and change management – is a body of KT activities intended to enhance employee skill sets through general education, competency-based instruction, and task-specific training. In many organizations, T&D expertise resides in human resources (HR) departments. However, T&D responsibilities and activities are often dispersed throughout organizations independent of HR departments. External resources, including education institutions, professional organizations, and contractors often provide these independent T&D activities. See *Organizational Development* and *Change Management.*

Transformational Leadership At its most basic level, transformational leadership in organizational management means helping individuals and groups progress from a current level of performance to an improved level. It focuses on the "Four Is": *idealized influence, inspirational motivation, intellectual stimulation*, and *individual consideration*. See *Primary Functions of Organizational Management.*

Triple Bottom Line The triple bottom line is a true-cost accounting framework attributed to sustainability thought leader John Elkington. It has three parts: social, environmental, and economic, aka "people, planet, and profit" (3Ps). Its purpose is to measure and responsibly respond to an organization's social, environmental, and financial performance. See *The Prime Directive of Finance.*

Value Chain A value chain is the processes or activities by which a company adds value to a product or service, including design, production, marketing, distribution, and after-sales support. *See Supply Chain.*

Value Creation Value creation in business is the process of converting inputs into outputs that are worth more than their components.

Value Engineering Value engineering is a systematic, organized approach to providing necessary functions in a process at the lowest cost. It promotes substituting materials and methods with less expensive alternatives without sacrificing functionality or quality to increase the value of the product.

Variances Variances occur when a requirement, standard, or plan is followed, yet the work effort results differ from what is planned and expected. Although such results are typically negative, they can be positive. When the results are negative, actions must be taken to control or correct them to avoid consequences. When the results are positive, they should be highlighted as lessons learned in the periodic project-progress reports, the end-of-project closure reports, or management systems reviews. See *Nonconformances, Nonconformities, and Noncompliances* and *Conformance versus Compliance.*

Venn Diagram A Venn diagram illustrates relationships between concepts using overlapping circles. The amount of overlap represents the strength of such relationships.

Window Dressing Window dressing is the act of making something appear deceptively attractive or favorable. Greenwashing is a form of window dressing. See *Greenwashing.*

Work Process A work process is the systematic spatial and chronological sequence of activities involving people, techniques, materials, energy, and information. It has three principal phases: inputs, input-transforming activities, and outputs and outcomes.

Biographies

William Borges, BA, MBA

As a transformational sustainability leader, Bill helps organizations systematically identify and address their most pressing environmental, social, and business needs within their resource limits and opportunities using continuous improvement (CI) methods.

He has directed and participated in over 100 sustainability and environmental management projects. These included the landmark Denver Metropolitan Region Clean Air and Clean Water Plans recognized for innovation by the President's Council on Environmental Quality. As a principal environmental scientist, he initiated an environmental management system (EMS) practice at Midwest Research Institute (MRI), now MRI Global, co-operator of the United States Department of Energy's National Renewable Energy Laboratory. Earlier, he designed and helped implement United States Marine Corps Base Camp Pendleton's highly successful EMS, one of the first at a United States Department of Defense facility.

His organizational management experience includes reorganizing operations and sales departments of an eventual Malcolm Baldrige National Quality Award winner, helping grow a regional environmental engineering firm as the chief administration officer, restructuring firms supported by a small venture capital group, and introducing a CI management system (MS) to a regional integrated healthcare delivery system.

In addition to teaching undergraduate and graduate business management courses, Bill designed and taught advanced professional certificate programs and courses, including *The Sustainable Organization, Environmental Management Systems,* and *Sustainable Supply Chain Management*.

His volunteer work includes chairmanship of the *Orange County Sustainability Collaborative*; examiner for the *Nevada APEX Awards*, a state-level *Malcolm Baldrige National Quality Award* program; a Grand Award judge in

Sustainability Programs: A Design Guide to Achieving Financial, Social, and Environmental Performance, First Edition. William Borges and John Grosskopf.

the environmental management and engineering categories at the *Regeneron International Science and Engineering Fair*; and a scholar selection committee member and mentor for the *Tiger Woods Foundation*. He also participated in the *Sustainability Accounting Standards Board's Healthcare Industry Advisory Group*.

His publications include *Planning the Creation and Implementation of World Standard Environmental Management Systems*, published jointly by MRI and the Korean Chamber of Commerce and Industry; a blog, *Creating Hospital Sustainability Programs*; and *Performance Improvement for Healthcare Managers*, published by Renown Health.

Bill holds a Bachelor of Arts in Geography with an emphasis on environmental management and a Master of Business Administration with a thesis on designing, implementing, and managing corporate EMSs. Before completing his bachelor's degree, he served as a special investigations case controller with the United States Army Intelligence Command.

He is enjoying a well-deserved retirement with his family in a small city between the Phoenix and Tucson metropolitan areas.

LinkedIn profile: https://www.linkedin.com/in/william-borges-7a62876
wm.borges.3@gmail.com

John Grosskopf, BSCE, PE, BCEE (ret.)

For more than four decades, John has helped hundreds of public and private organizations dramatically improve sustainability, environmental, health, and safety (EHS), and security performance while substantially reducing costs, risks, and impacts. His pioneering work led to several firsts and significant achievements in environmental management by developing and implementing specialty CI MSs. His sustainability work began in the 1990s – long before it became a familiar concept – and continues today.

He helped lead General Dynamic's highly acclaimed environmental programs as a chief architect of its pioneering EMS. His leadership positions included General Dynamics' first Corporate Environmental Manager, President of the Rocky Mountain Association of Environmental Professionals, Chairman of the California Aerospace Environmental Association, and Founder and first Chairman of the Orange County Sustainability Collaborative.

His notable program-innovation achievements include:

- Practicable zero-emissions, zero-hazardous waste, zero-discharge, and total water recycling during a fourfold increase in production with no EHS sanctions at a major defense production facility in stringently regulated southern California

- A principal author and pilot program test leader for the Israeli Security Management System, the world's first national security management system and predecessor to the ISO 27001 Information Management System standard
- Chief designer of and first to pilot test the United States Environmental Protection Agency's (USEPA) compliance-focused EMS model still in use for compliance settlement decrees and programs, and
- Principal consultant to the United States National Park Service, for which he reduced EHS hazards and impacts at Grand Teton National Park, Hot Springs National Park, Curecanti National Recreation Area, Glen Canyon National Recreation Area, and others.

John has educated, instructed, and trained hundreds of individuals, published extensively, and spoken frequently on advanced EHS, ISO management systems, security, and sustainability topics throughout the United States and abroad. He was an adjunct faculty member at the University of California (UC) San Diego. He also lectured at UC Irvine, San Diego State University, the University of Colorado at Boulder, Washington State University, and others.

His work earned numerous awards and distinctions from the USEPA, the United States National Park Service, the American Society of Civil Engineers, the South Coast Air Quality Management District, and others.

He holds a Bachelor of Science in Civil Engineering. He has held Professional Engineer (PE) licenses in Illinois and Colorado and is a retired Board Certified Environmental Engineer through the American Academy of Environmental Engineers and Scientists.

John lives in the Dallas-Fort Worth area, where he consults, writes, and volunteers for local communities. He enjoys traveling with his wife, hiking, biking, boating, and astronomy.

LinkedIn profile: https://www.linkedin.com/in/johnwgrosskopf
jwgrosskopf@gmail.com

References

Articles

Abbosh, O., Shim, C., Goos, E., et al., 29 February 2024, Beyond Checking the Box: How to Create Business Value with Embedded Sustainability, *IBM Institute for Business Value, Research Insights*, Retrieved 27 March 2024: www.ibm.com/thought-leadership/institute-business-value/en-us/report/sustainability-business-value

Birch, K., 13 February 2024, Deloitte Leads Way on Employee Upskilling in Sustainability, *Sustainability Magazine*, Retrieved 1 April 2024: sustainabilitymag.com/sustainability/deloitte-leads-way-for-employee-upskilling-in-sustainability

Cohen, S., 11 December 2023, The Importance of Sustainability Metrics to Sustainability Management, *Columbia University Climate School*, Retrieved 14 December 2023: news.climate.columbia.edu/2023/12/11/the-importance-of-sustainability-metrics-to-sustainability-management/

Cote, C., 13 April 2021, Making The Business Case for Sustainability, *Harvard Business School Online Business Insights Blog*, Retrieved 12 March 2024: online.hbs.edu/blog/post/business-case-for-sustainability

DeSmet, A., Mugayar-Baldocchi, M., Reich, A, et al., 11 September 2023, Some Employees Are Destroying Value. Others Are Building It. Do You Know the Difference? *McKinsey Quarterly*, *McKinsey & Company*, Retrieved 9 September 2023: https://www.mckinsey.com/capabilities/people-and-organizational-performance/our-insights/some-employees-are-destroying-value-others-are-building-it-do-you-know-the-difference

Doherty, R., Kampel, C., Koivuniemi, A., et al., 9 August 2023, The Triple Play: Growth, Profit, and Sustainability, *McKinsey & Company,* Retrieved 25 March 2024: www.mckinsey.com/capabilities/strategy-and-corporate-finance/our-insights/the-triple-play-growth-profit-and-sustainability

Eapen, S., 2 August 2017, How to Build Effective Sustainability Governance Structures, *BSR Sustainable Business Network and Consultancy, Blog Post*, Retrieved

28 February 2024: www.bsr.org/en/blog/how-to-build-effective-sustainability-governance-structures

Eccles, R.G. and Taylor, A., July–August 2023, The Evolving Role of Chief Sustainability Officers, *Harvard Business Review*, Retrieved 9 July 2023: hbr.org/2023/07/the-evolving-role-of-chief-sustainability-officers

Faelli F., Lichtenau, T., Blasberg, J., et al., 25 October 2023, The Visionary CEO's Guide to Sustainability, *Bain & Company*, Retrieved 29 March 2024: www.bain.com/insights/topics/ceo-sustainability-guide/

Greco, L. and Silverman, Z., 4 May 2023, Ready, Set, Go, and Keep Going: Why Speed Is Key to a Successful Transformation, *McKinsey & Company*, Retrieved: 6 June 2024: www.mckinsey.com/capabilities/transformation/our-insights/ready-set-go-and-keep-going-why-speed-is-key-for-a-successful-transformation

Isfordink, P. and Tan, V., 17 July 2023, Why 'Change Management' Is Essential for Your Sustainable Ambitions: Three Pillars: From Sustainability Ambitions to Sustainable Results, *PWC Netherlands*, Retrieved 1 April 2024: www.pwc.nl/en/topics/blogs/why-change-management-is-essential-for-your-sustainable-ambitions.html

Kenlon, T., 18 May 2023, Need Sustainability Strategy Training? Look Here, GreenBiz.com, Retrieved 20 May 2023: www.greenbiz.com/article/need-sustainability-strategy-training-look-here

Kenlon, T., 3 January 2023, Which of These 40 Sustainability Certifications Is Right for You?, GreenBiz.com, Retrieved 20 May 2023: www.greenbiz.com/article/40-sustainability-certifications

Korkmaz, B., Nuttall, R., Pérez, L., et al, 26 May 2023, ESG Momentum: Seven Reported Traits That Set Organizations Apart, *McKinsey & Company*, Retrieved 28 March 2024: www.mckinsey.com/capabilities/strategy-and-corporate-finance/our-insights/esg-momentum-seven-reported-traits-that-set-organizations-apart

Makower, J., 19 July 2021, Inside the War for ESG Talent, GreenBiz.com, : www.greenbiz.com/article/inside-war-esg-talent

McIntire, L., 11 March 2015, Ten Characteristics of the World's Best Corporate Sustainability Programs, *CSRWire Blogs*, *Triple Pundit Newsletters*, Retrieved 22 May 2024: www.triplepundit.com/story/2015/ten-characteristics-worlds-best-corporate-sustainability-programs/87481

McNeive, A., 17 January 2024, Change Management for Sustainability, *PROSCI, Inc.*, Retrieved 1 April 2024: www.prosci.com/blog/change-management-for-sustainability

Rade, A. ed, 27 March 2023, 12 Steps to Start Your Corporate Sustainability Program: A Guide to Help You Get Your Company's Sustainability Program Off The Ground, *Workiva/Sustain.Life*, Retrieved 11 March 2024: www.sustain.life/blog/12-steps-corporate-sustainability-program

Winston, A., October 3, 2023, The Burden of Proof for Corporate Sustainability Is Too High, *MIT Sloan Management Review*, Retrieved 27 March 2024: sloanreview.mit .edu/article/the-burden-of-proof-for-corporate-sustainability-is-too-high/

Young, D. and Gerard, M., 9 April 2021, How to Tell if Your Business Model Is Truly Sustainable, *BCG Henderson Institute, Boston Consulting Group*, Retrieved 22 May 2024: www.bcg.com/publications/2021/nine-attributes-to-a-sustainable-business-model

Books

Mohin, T.J., Changing Business from the Inside Out: A Treehugger's Guide to Working in Corporations, 6 August 2012, Berrett-Koehler Publishers, Oakland, California, ISBN 978-1609946401.

Pfeiffer, J., The Human Equation: Building Profits by Putting People First, 1998, Harvard Business School Press, ISBN 9780875848419.

Russell, J.P., ed, The Quality Audit Handbook, American Society for Quality Press Audit Division, 1 January 2000, McGraw-Hill, ISBN 978-0873894609.

Willard, B., The Sustainability Champion's Guidebook: How to Transform Your Company, 2009, New Society Publishers, Gabrioloa Island, British Columbia, Canada, ISBN 978-0-86571-658-2.

Willig, J.T., ed, Auditing for Environmental Quality Leadership, 17 April 1995, Wiley, New York, NY, ISBN 0-471-11492-8.

Case Study Sources

Grosskopf, J. and Borges, W., Perdue Farms Child Labor Case: Subcontractor's Alleged Violations of Child Labor Laws, Retrieved 29 May 2024:

- https://www.nytimes.com/2023/09/23/us/tyson-perdue-child-labor.html
- news.bloomberglaw.com/daily-labor-report/perdue-tyson-foods-face-unique-probe-in-child-labor-crackdown
- https://www.dol.gov/agencies/whd/child-labor
- www.jurist.org/commentary/2023/12/child-labor-investigation-at-tyson-foods-inc-is-supply-chain-due-diligence-the-next-step/
- www.hklaw.com/en/news/intheheadlines/2023/10/perdue-tyson-foods-face-unique-probe-in-child-labor-crackdown

Borges, W., The Current State of Advanced Information Technology Applications for Sustainability, Retrieved 8 May 2024:

- www.mckinsey.com/capabilities/mckinsey-digital/our-insights/moving-past-gen-ais-honeymoon-phase-seven-hard-truths-for-cios-to-get-from-pilot-to-scale

- www.forbes.com/sites/forbestechcouncil/2023/11/22/14-ways-ai-can-help-business-and-industry-boost-sustainability/?sh=32048b86ed04
- hbr.org/2023/10/the-opportunities-at-the-intersection-of-ai-sustainability-and-project-management
- www.wiley.com/en-us/Artificial+Intelligence+for+Sustainable+Applications-p-9781394175239
- www.pwc.co.uk/sustainability-climate-change/assets/pdf/how-ai-can-enable-a-sustainable-future.pdf
- research.aimultiple.com/sustainability-ai/
- www.sciencedirect.com/science/article/abs/pii/S0040162522003262

Grosskopf, J. and Borges, W., The Teck Resources Limited Red Dog Mine Management Review Processes

- Teck 2023 Sustainability Report, 11 March 2024, *Teck Resources Limited*, Retrieved 24 June 2024: www.teck.com/media/2023-Sustainability-Report.pdf
- Questionnaire Provided by the Authors to and Completed by S. Staley, et al, *Teck Resources*, July 2024.
- Grosskopf/Orion ISO 14001 RDM verification audits 2019, 2020, 2021

Grosskopf, J. and Borges, W., The Volkswagen Emissions Scandal, Retrieved 27 March 2024:

- www.statista.com/statistics/466109/annual-closing-share-prices-of-volkswagen/
- qz.com/volkswagen-dieselgate-scandal-rupert-stadler-audi-1850581201
- www.reuters.com/article/idUSKBN2141JA/
- www.justice.gov/opa/pr/volkswagen-spend-147-billion-settle-allegations-cheating-emissions-tests-and-deceiving
- knowledge.wharton.upenn.edu/podcast/knowledge-at-wharton-podcast/volkswagen-diesel-scandal/

Corporate ESG and Sustainability Reports

2022 Citizenship Report, *Mattel, Inc.*, 18 September 2023, Retrieved 18 March 2024: assets.contentstack.io/v3/assets/bltc12136c3b9f23503/blt9629cb310f0aedf5/650b6a914b35c8d0fdfffe76/Mattel_Citizenship_Report_FINAL.pdf

2022 Environmental Sustainability Report, *Microsoft Corporation*, 31 May 31 2023, Retrieved 18 March 2024: query.prod.cms.rt.microsoft.com/cms/api/am/binary/RW15mgm

2022 Sustainability Report, *GE HealthCare*, Retrieved 20 September 2023: www
.gehealthcare.com/-/jssmedia/gehc/us/files/about-us/sustainability/reports/ge-
healthcare-sustainability-report-2022.pdf

Environmental, Social, and Governance Highlights FY2023 and Various Corporate
Website Pages, Walmart, Retrieved 23 February 2024:

- corporate.walmart.com/content/dam/corporate/documents/esgreport/fy2023-
  walmart-esg-highlights.pdf
- corporate.walmart.com/purpose/esgreport

Health for Humanity 2025 Goals, *Johnson & Johnson*, Retrieved 9 April 2024:

- www.jnj.com/about-jnj/policies-and-positions/priority-topics-assessment
- www.jnj.com/about-jnj/policies-and-positions
- www.jnj.com/health-for-humanity-goals-2025
- www.jnj.com/about-jnj/policies-and-positions/our-position-on-environmental-
  health-and-safety-management

Sustainable Aerospace Together: 2023 Sustainability Report, *The Boeing Company*,
21 December 2023, Retrieved 1 April 2024: www.boeing.com/sustainability/
annual-report

Task Force on Climate-Related Financial Disclosures Report, *Verizon*, 9 September
2021, Retrieved 11 November 2023: www.verizon.com/about/sites/default/files/
Verizon-2021-TCFD-Report.pdf

Other Resources

Brundtland, G.H., ed, Report of the World Commission on Environment and
Development: Our Common Future, *United Nations*, 1987, Retrieved 22 February
2024: sustainabledevelopment.un.org/content/documents/5987our-common-
future.pdf

Business Policy - Definition and Features, *Managementstudyguide.Com*, Retrieved
3 March 2024: managementstudyguide.com/business-policy.htm

Global Green Skills Report 2023, 9 June 2023, LinkedInc.com, Retrieved 20 May 2024:
economicgraph.linkedin.com/content/dam/me/economicgraph/en-us/global-
green-skills-report/green-skills-report-2023.pdf

Guidance: Making Environmental Claims on Goods and Services, *Competition and
Markets Authority, Government of the United Kingdom*, 20 September 2021,
Retrieved 25 March 2024: www.gov.uk/government/publications/green-claims-
code-making-environmental-claims/environmental-claims-on-goods-and-services

ISO Handbook: The Integrated Use of Management System Standards, *The International Organization for Standardization*, 2nd Ed, November 2018, Retrieved 14 May 2024: https://www.iso.org/files/live/sites/isoorg/files/store/en/ PUB100435_preview.pdf

Microsoft and Boston Consulting Group, Nov 2, 2022, Closing the Sustainability Skills Gap: Helping Businesses Move from Pledges to Progress, *Microsoft Corporation*, Retrieved 20 May 2023: query.prod.cms.rt.microsoft.com/cms/api/am/binary/ RE5bhuF

Index

Sustainability Programs: A Design Guide to Achieving Financial, Social, and Environmental Performance,
First Edition. William Borges and John Grosskopf.
© 2025 John Wiley & Sons, Inc. Published 2025 by John Wiley & Sons, Inc.